BIOSEPARATION ENGINEERING

Progress in Biotechnology

Volume 1 New Approaches to Research on Cereal Carbohydrates (Hill and Munck, Editors)

Volume 2 Biology of Anaerobic Bacteria (Dubourguier et al., Editors)

Volume 3 Modifications and Applications of Industrial Polysaccharides (Yalpani, Editor)

Volume 4 Interbiotech '87. Enzyme Technologies (Blažej and Zemek, Editors)

Volume 5 In Vitro Immunization in Hybridoma Technology (Borrebaeck, Editor)

Volume 6 Interbiotech '89. Mathematical Modelling in Biotechnology (Blažej and Ottová, Editors)

Volume 7 Xylans and Xylanases (Visser et al., Editors)

Volume 8 Biocatalysis in Non-Conventional Media (Tramper et al.,Editors)

Volume 9 ECB6: Proceedings of the 6th European Congress on Biotechnology (Alberghina et al., Editors)

Volume 10 Carbohydrate Bioengineering (Petersen et al., Editors)

Volume 11 Immobilized Cells: Basics and Applications (Wijffels et al., Editors)

Volume 12 Enzymes for Carbohydrate Engineering (Kwan-Hwa Park et al., Editors)

Volume 13 High Pressure Bioscience and Biotechnology (Hayashi and Balny, Editors)

Volume 14 Pectins and Pectinases (Visser and Voragen, Editors)

Volume 15 Stability and Stabilization of Biocatalysts (Ballesteros et al., Editors)

Volume 16 Bioseparation Engineering (Endo et al., Editors)

Supported by

The Commemorative Association for the Japan World Exposition (1970)

Progress in Biotechnology 16

BIOSEPARATION ENGINEERING

Proceedings of an International Conference on Bioseparation Engineering: "Recovery and Recycle of Resources to Protect the Global Environment", organized under the Special Research Group on Bioseparation Engineering in the Society of Chemical Engineers, Japan
Nikko, Japan, July 4-7, 1999

Edited by

I. Endo
Biochemical Systems Laboratory, RIKEN Institute,
Saitama, Japan

T. Nagamune
Department of Chemistry & Biotechnology, University of Tokyo,
Tokyo, Japan

S. Katoh
Department of Chemical Science and Engineering, Kobe University,
Kobe, Japan

T. Yonemoto
Department of Chemical Engineering of Tohoku,
Sendai, Japan

2000

ELSEVIER
Amsterdam – Lausanne – New York – Oxford – Shannon – Singapore - Tokyo

ELSEVIER SCIENCE B.V.
Sara Burgerhartstraat 25
P.O. Box 211, 1000 AE Amsterdam, The Netherlands

First edition 2000

Library of Congress Cataloging in Publication Data
A catalog record from the Library of Congress has been applied for.

ISBN: 0 444 50441 9

Transferred to digital printing 2006

Preface

Bioseparation process systems are most influential upon quality and quantity of the products of the bioindustry. The process systems, therefore, determine stability, safety and cost of the bioproduct.

The process systems consist of various unit operations like centrifugation, precipitation, chromatography, membrane separation, crystallization and so on. These operations are executed in special order according to the product. The characteristic features of this process system are summarized as follows:
1)The product is contained in the culture broth at a low concentration and in a complete mixture with many other compounds.
2) The product material is very sensitive to temperature, pressure, pH and to other operation variables.
3) The bioproduct is required to be of high quality in activity and / or in purity. The production is restricted by certain laws and regulations. Hereby, the bioseparation process should often be operated under mild condition in the clean room which is determined by regulation.

Recently, regulations in terms of environment protection became common in the world. Bioindustries in any countries can not neglect this social pressure. In other words, close to zero emission from factory is strongly requested particularly in advanced countries like U.S.A., the EC countries and Japan. Thus, bioseparation engineering of today is going to include downstream process engineering such as waste water, material and gas treatment. Taking into account this tendency in the world, we, bioseparation process engineers in Japan who gathered to the special research group on bioseparation engineering in the Society of Chemical Engineers, Japan planned the international conference on bioseparation engineering at Nikko, Japan during July 4th to 7th under the main theme of "Recovery and Recycle of Resources to Protect the Global Environment ".

The scope of this book, is based on the conference, and deals with not only the recent advances in bioseparation engineering in a narrow sense but also the environmental engineering which includes waste water treatment and bioremediation The contributors of this book cover many disciplines, including such as chemical engineering, analytical chemistry, biochemistry, microbiology and so on.

This book contains the following 5 chapters:

The editors do hope strongly that the content of this book would stimulate young engineers and scientists who will develop the bioseparation engineering further in 21C. and contribute to a world-wide attention to the global environment.

We thank Professors Sven-Olof Enfors (Royal Institute of Technology, Sweden), Michael R. Ladish (Purdue University, U.S.A.) and Rainer Rudolph (Martin-Luther University, Germany), for their valuable contribution to the review of manuscripts in this book.

The Editors, I. Endo, T. Nagamune, S. Katoh and T. Yonemoto

Acknowledgments

The Organizing Committee gratefully acknowledges the support of the followings sponsors:

- Amersham Pharmacia Biotech AB.
- Ajinomoto Co., Inc.
- Asahi Chemical Industry Co., Ltd.
- Japan Bioindustry Association
- Japan Society for Promotion of Science
- Kaneka Co.
- Kirin Brewery Co., Ltd.
- Nihon Millipore Ltd.
- Mitsui Chemicals Inc.
- Osaka Pharmaceutical Manufactures Association
- Special Research Group on Bioseparation Engineering,
  The Society of Chemical Engineers, Japan
- The Commemorative Association for the Japan World Exposition (1970)
- The Japan Research Institute, Ltd.
- The Pharmaceutical Manufactures Association of Tokyo

Contents

Chapter 1

Adsorption, Chromatography and Membrane Separations

Bioseparation Engineering
I. Endo, T. Nagamune, S. Katoh and T. Yonemoto (Editors)

Recent Advances in Membrane Technology that Could Improve Resource Recovery and Recycle: Fluid Mechanics, Surface Science and Bioaffinity

Georges Belfort

Howard P. Isermann Department of Chemical Engineering
Rensselaer Polytechnic Institute, Troy, NY 12180 (USA)

THE GLOBAL ENVIRONMENT

With the realization that enormous investments will be needed to balance economic activity with environmental protection (called sustainable development), new clean and cleaning technologies will be needed to address the global conditions of excessive pollution, increasing population and increasing industrialization[1]. Of these technologies,

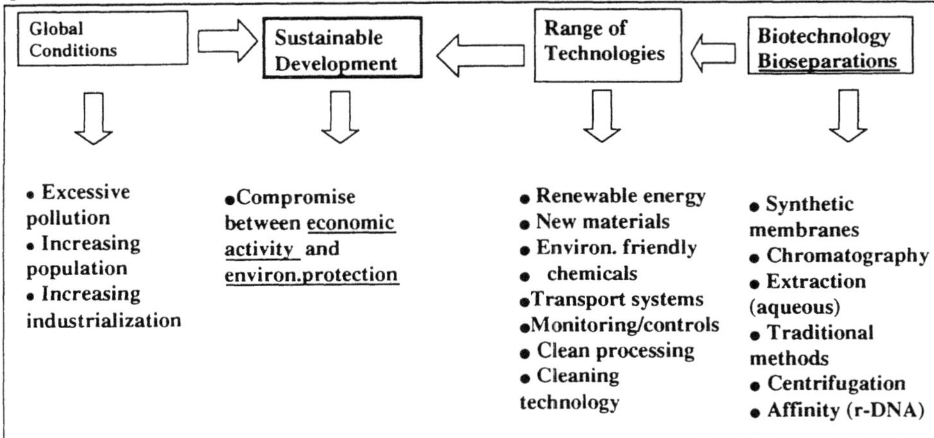

Fig. 1 Sustainable development how can synthetic membrane technology contribute? Ref: B. Zechendorf, TIBTECH 17, 219 (1999)

synthetic membrane technology is expected to be a major player. See **Fig. 1**. The reasons for this are that pressure-driven membrane processes are very attractive because they do not involve a phase change (i.e. do not consume large amounts of energy), are often linearly scalable, do not need additives, are relatively fast (rate governed rather than equilibrium processes), operate in a continuous mode, are easily combined with other processes, and are completely contained. However, several limitations, still need to be addressed. Foremost among these are concentration polarization (CP) and fouling phenomena which can substantially reduce performance through osmotic effects and solute adsorption and deposition on the membrane surface. These limitations can readily result in additional energy requirements and larger capital and maintenance costs, thus

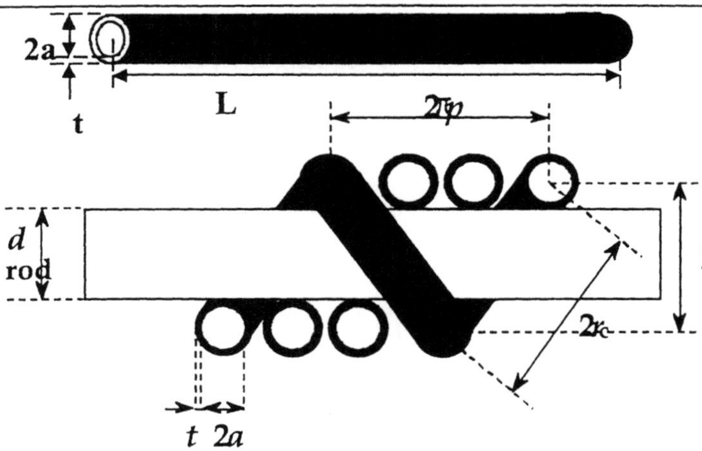

Fig. 2. Schematic of (a) the current commercial linear hollow fiber and (b) a new helical hollow fiber in which centrifugal vortices are produces that clean the membrane surface.

reducing the attractiveness of pressure-driven membrane technology. Various approaches have been used to address these limitations including improved methods of operation through the use of positive displacement pumps for controlling permeation rate and minimizing transmembrane pressure drop, operating at or below a prescribed protein wall concentration, modifying the chemical properties of the membrane surface so as to minimize solute-membrane interactions, and improved fluid mechanics and module design for reducing solute concentration and deposition on the membrane.

SYNTHETIC MEMBRANE TECHNOLOGY

The success of synthetic membrane technology has depended on a collaboration between polymer and surface scientists, who have developed suitable commercial membranes, and chemical engineers with an expertise in mass transfer and fluid mechanics, who have designed modules for optimizing filtration performance. Recent developments in these two fields will be emphasized in this presentation with a particular focus on biotechnology and the need to recover valuable proteins from solution. We argue that the need to understand the behavior of fluid flow with imposed centrifugal vortices can assist in designing optimal flow paths with minimal fouling and reduced concentration polarization[2,3]. Similarly, the connection between a fundamental understanding of intermolecular forces between a model protein, hen egg lysozyme (Lz), and polymeric membranes is crucial for the development of new and improved membrane materials for this application[4,5].

THREE FUNDAMENTAL EXAMPLES

An example of the first module design without moving parts especially designed for suspensions commonly found in the biotechnology industry is our new "Da Vinci" module. By flowing sufficiently fast along a helical twisted membrane tube, counter rotating Dean vortices can be used to clean the membrane surface and reduce particulate build-up and fouling. See Fig. 2.

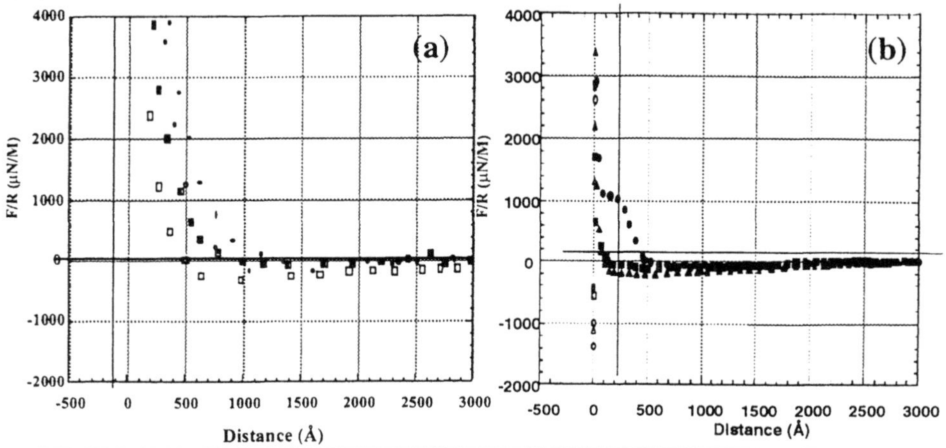

Fig. 3 Protein-membrane force-distance curves for (a) a HEMA-modified poly(ethersulfone) membrane and (b) an unmodified poly(ethersulfone) membrane. The feed contained hen egg white lysozyme at 50 mg/l and pH 6.6

Koehler et al.[4,5] have explained the well-known phenomenon of increased protein fouling on hydrophobic (poly(sulfone), PES) as compared to hydrophilic (hydroxyethyl methacrylate-PES, HEMA/PES) surfaces by using a correlation between adhesion forces and filtration fluxes. See **Fig.** 3. They show that protein-protein and protein-polymer interactions are about equally important for the PES-Lz system, while only protein-polymer interactions are important for the HEMA/PES-Lz system. How these two surfaces effect the stability of Lz and the fouling of membranes is discussed in detail.

Synthetic membranes or porous chromatographic beads are attractive binding media for affinity separations of fusion proteins because they overcome diffusion limitations with convective flow. In our final example, we illustrate the development and application of a new linker with controllable cleavage activity between the binding domain and the desired protein[6]. See **Fig.** 4. Both batch and column examples of the resulting one-step purification using temperature and pH excursions to induce cleavage are presented. Excellent purity and yield are obtained in all cases.

CONCLUSIONS

Cost estimates for achieving sustainable development up to the year 2,000 are about twice the current world pharmaceutical market of US$308 billion![7,8]. Whether the advanced societies will be prepared to spend such a large amount without a crisis or environmental disaster, is open to question. Clearly, attractive technologies that utilize less energy and produce less waste such as biotechnology and synthetic membrane processes are prime candidates for such an effort.

ACKNOWLEDGEMENTS

The author thanks his past and current graduate students, post-docs and research collaborators. Technical support was obtained from Millipore Corp., Bedford, MA., while funding was supplied by Bob Peterson, Dow Chemical Co. and FilmTec Corp., NWRI, NSF (CTS-9400610), DOE (DE-FG02-90ER1414)Millipore Corp., and the NATO Scientific Committee.

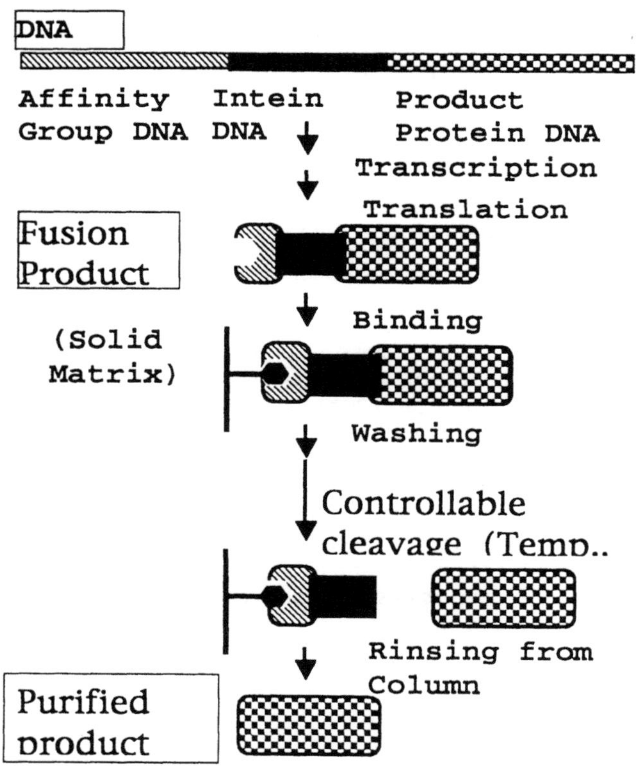

Fig. 5 Development of a new controllable linker for fusion protein

REFERENCES

1. B. Zechendorf, Trends in Biotechn. 17, (1999) 219-225.
2. G. Gehlert, S. Luque, and G. Belfort, Biotechnology Progress, 14, (1998) 931-942. .
3. S. Luque, H. Mallubhotla, G. Gehlert, R. Kuriyel, S. Dzengeleski, S. Pearl, and G. Belfort, Biotechnology Bioengineering., (1999) in press.
4. J. A, Koehler, M. Ubricht and G. Belfort, Langmuir 13, (1997) 4162.
5. J. A, Koehler, M. Ubricht and G. Belfort. Langmuir, (1999) in review.
6. D. Wood, W. Wei, V. Derbyshire, G. Belfort, and M. Belfort, Nature Biotechnology, (1999) in press.
7. J. MacNeil, Scientific Amer. (1989)105-113.
8. S. Walker, Plenary lecture at the "Recovery of Biological Products IX", Whistler, Canada, May 23, 1999.

Bioseparation Engineering
I. Endo, T. Nagamune, S. Katoh and T. Yonemoto (Editors)

Stabilization of target protein during bioseparation

X.-L. Feng[a], Y.-T. Jin[b] and Z.-G. Su[a]

[a]National Laboratory of Biochemical Engineering, Institute of Chemical Metallurgy, Chinese Academy of Science, Beijing 100080, The People's Republic of China

[b]Laboratory of Biochemical Engineering, Dalian University of Technology, Dalian 116012, The People's Republic of China

Denaturation of target protein by various separation and purification steps contributes significant part to the total product loss in bioseparation. This report classifies the denaturation into four types including thermal denaturation, shear denaturation, solution denaturation and adsorption denaturation. For stabilization of target protein, three strategies are proposed including careful selection of unit operation to avoid detrimental action, process optimization to reduce the number of steps and the total processing time, and utilization of protective reagents such as PEG during bioseparation. It is important to understand the structure and property of the product to design the best bioseparation route.

1. INTRODUCTION

Low recovery is a major problem in production of pharmaceutical proteins. The loss of target protein can be classified into two aspects. The first one is physical loss in the flow stream, such as the leakage through an ultrafiltration membrane during concentration operation, the carry-away during a washing step in chromatography after loading, or even the residual left in the dead volume of a process device and the pipelines. This part of loss should not contribute to more than 15%, and is often controllable by proper process design and operation. The second loss is the denaturation of the target protein by various separation or purification steps. This part is significant, much more than 15%, and is difficult to control.

Any separation step in a bioprocess relies on its physical, chemical or biological action to distinct one or a group of proteins from the other. The product, or the target protein, has a limited stability undergoing the treatment. Even there is no change in the molecular weight or in the one dimensional structure, a minor alteration of the molecular conformation would result in loss of its biological activity. While molecular biologists are trying to construct artificial proteins that are more stable and functional, biochemical engineers are working hard in designing optimal separation routes to maintain the three dimensional integrity of the products and to achieve the desired purification during bioseparation [1].

---

* This research is supported by China Natural Science Foundation, Grand No. 29525609 and 29736180

Table 1
Denaturation of proteins in separation and purification

Unit operation	Separation principles	Damage to proteins
Cell disruption	Liquid shear, impingement, pressure change, hydrolysis of cell membrane & wall	Thermal denaturation, shear denaturation, solution denaturation
Aqueous two-phase extraction	Partition in different phases driven by thermodynamics	Solution denaturation, shear denaturation
Centrifugation	Density difference	Thermal denaturation
Membrane filtration	Size difference	Shear denaturation, adsorption denaturation
Chromatography	Surface interaction, size difference	Adsorption denaturation, solution denaturation
Freeze drying	Volatility difference	solution denaturation

2. AVOIDANCE OF DETRIMENTAL ACTION

In order to decrease the denaturation loss, care has to be exercised in choosing suitable separation methods to avoid detrimental actions, such as increasing temperature, excessive stirring, marked changes in pH , adding organic solvents and exposure to ultraviolet light. Table 1 lists the frequently used unit operations, its separation principles and possible damage to proteins. In general, protein denaturation in bioseparation can be classified into four categories, i.e. thermal denaturation, shear denaturation, solution denaturation and adsorption denaturation. Other denaturations such as those induced by high pressure and ultraviolet light are not common, and will not be discussed here.

Thermal denaturation is caused by temperature increase, resulting in disorder of the three dimensional structure by breakage of the forces stabilizing the spatial conformation, such as hydrogen bonds, electrostatic and hydrophobic interactions. In mechanical cell disintegration such as homogenization and bead milling, part of the mechanical energy transferred to heat energy, increasing the temperature of the homogenate. For example, one passage through a homogenizer at 600 bars can increase the homogenate temperature by 2-5 ℃ depending on cell concentration and viscosity of the homogenate. Cooling is necessary for multiple passage of homogenization.

Shear denaturation is associated with high liquid flow rates. The mechanism is still unclear. Many observations have proved that protein may lost its activity in a high liquid shear field.

For shear sensitive proteins, cross-flow microfiltration and ultrafiltration may cause denaturation due to high shear used for minimization of concentration polarization. Pumping is a process associated with liquid shear. Peristaltic pumps are normally regarded as mild operators and preferred choice for less contamination. However, studies have demonstrated that peristaltic pumps could denature proteins by generation of protein aggregates. The solution of serum albumin, in which aggregates had been removed, when being pumped again with a peristaltic pump, produced aggregates again. The pumping period and concentration of the protein determine the magnitude of aggregate formation [2].

For solution denaturation, several mechanisms may be involved, including protease hydrolysis, chemical hydrolysis, interaction with salts, surfactants, organic solvents etc.[3]. In fact these actions in solution may be going on all the time during bioseparation with varied degrees for different proteins, even the solution is in cold storage. When a separation requires addition of certain substances to the protein solution and process it under certain condition, denaturation by the substances present in the solution may occur. For example, chemical disruption of the cells requires addition of organic solvents, surfactants or chaotropic agents such as guanidine hydrochloride. These reagents break down cell membranes to release the intracellular protein. However, the released product is also under the attack of the reagents. Aqueous two-phase extraction in general is good for maintaining the activity of the protein, but the high concentration of salts and type of salts may affect the protein activity in salt-polymer system. Solution denaturation depends on the concentration of the solutes that denature the product. In freeze drying, much of the protein activity may be lost during freezing stage because water forms ice and solute concentrations are increased.

Adsorption denaturation happens on solid surface. Non-specific adsorption of a protein to the surface of a separation medium or any contacting materials of the process contributes to the denaturation significantly. Specific adsorption is a basis of chromatographic separation. For purification of pharmaceutical proteins, chromatographic steps must be involved. However, most chromatographic media are not totally selective with uniform adsorption pattern. Protein denaturation may take place on the surface of chromatographic media. Furthermore, elution of the target protein from the column requires specific solutions, such as those with extreme pH, high salt concentration or detergents. Considerable denaturation may occur during elution, especially in the case of affinity chromatography where the protein binds the ligand tightly, and harsh elution condition must be employed.

The four types of denaturation may happen simultaneously and interact with each other. For example, increasing temperature could not only cause thermal denaturation but also promote solution denaturation. High liquid shear also increases the temperature of the solution.

3. PROCESS OPTIMIZATION

It is understandable that the less the processing time and steps, the less the protein denaturation could be. In fact the rate of protein denaturation varies with different steps of bioseparation. As a general rule, protein should be processed as fast as possible. Inactivation of certain enzymes was found to be an exponential function of time [4] as

$$C_{active} = C_o \times e^{-t/k} \tag{1}$$

where C_{active} is the remaining activity after time t, C_o is the original activity, and k is a coefficient related to the protein structure and environment. Therefore, reduction of processing time is an obvious strategy for increasing protein recovery. During the last few years, process integration and optimization have been paid much attention. The goal is to make the process simpler and faster. Existing processes may be the duplicates of the protocols from molecular biology laboratories where the recombinant proteins were developed. Much of the concern at that time was placed on cloning and expression. As long as the protein can be purified, recovery is not the top priority. Such bioseparation process may be tedious, time consuming and high cost. It is the task for biochemical engineers to develop optimized process. In fact biochemical engineers should join the research at early stage of the product development because, for pharmaceutical proteins, any later change of the process after authoritative approval such as FDA approval must be re-validated.

A specific concern is chromatography. Though it is an indispensable operation, chromatography is a slow operation in which adsorption denaturation and solution denaturation occur. Attempts can be made on the following aspects:

1) to integrate an efficient pretreatment step with chromatography so that a large quantity of impurities are removed before chromatographic purification, reducing the number of chromatographic steps. An example is the integration of salt precipitation with hydrophobic interaction chromatography. After precipitation of impurities, the high salt concentration can be used directly as the feed for hydrophobic interaction chromatography.

2) to optimize chromatography techniques for the best purification. Chromatography in fact is a tricky operation involving medium selection, buffer selection, elution strategy etc.. Proper selection can result in high recovery and purification at a given chromatographic step. For purification of pharmaceutical proteins, it is often needed to have two more chromatographic steps. In this case, different combination of chromatographic steps will give different purification and recovery.

3) to use "direct-through" chromatography when the product concentration is high in the stream, i.e., to let product flow directly through the column in the loading process and to adsorb only impurities by the gel.

To further explain the concept of "direct-through" chromatography, an example is shown in Figure 1. It is the purification of a chemically modified protein with pI6.2. The impurity is the unmodified, native protein with pI7.1. Ion exchange chromatography is used. The left column is filled with anion exchanger where the product is adsorbed at pH6.5 . The impurity, with pI greater than the pH, is not adsorbed, flowing through the column. This is a typical adsorption chromatography for the product. The right hand side is the replacement where cation exchange column is used instead of the anion exchange column. The product with pI 6.2, is able to pass through the column at pH 6.5. Unmodified protein with positive charge is retained. Because the product concentration in the feed is as high as 90%, the advantages for flow-through

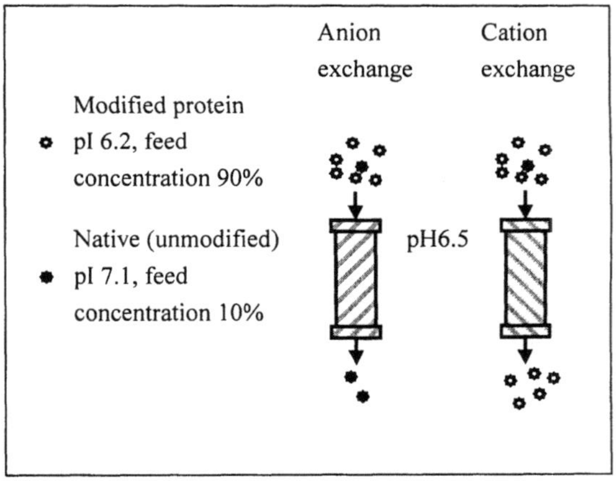

Figure 1 Comparison of "flow-through" and conventional adsorption chromatography for fractionation of chemically modified protein

chromatography can be viewed as

- reduction of process time, product going directly to the next step
- equipment (column, pump,etc) size reduction: up to 9/10 of the original
- chromatographic gel saving: up to 9/10 of the original
- no product denaturation due to adsorption & elution

4. USE OF PROTECTIVE REAGENTS

In Equation (1), the coefficient k is very important to determine the rate of deactivation. It varies with different proteins and solution environment. A large k indicates a stable protein at its stable environment. Increasing k value would slow down the rate of denaturation. The use of protective reagents in bioseparation is an effective way for protein stabilization. The known protective reagents include enzyme substrates or protein ligands, polyols such as glycerol, sucrose, specific salts and polymers.

Among the polymers, polyethylene glycol (PEG) is very useful. PEG has frequently been used for fractional precipitation of protein [5], for protein crystallization [6] and for aqueous two-phase separation [7]. Albertsson [8] had demonstrated that ovalbumin was easy to aggregate as soon as liquid shear was applied to the solution, and the aggregation could be prevented by addition of PEG.

About the effect of PEG on stability of proteins, there have been many reports on mechanism and application [9-10]. The earlier hypothesis of steric exclusion about the effect of PEG has been challenged by the mechanism of preferential exclusion [11].

There were several reports about that polyethylene glycol had the ability to increase protein partition coefficients in chromatographic processes, such as in size-exclusion chromatography, ion-exchange chromatography and protein A affinity chromatography [12-14]. The magnitude of the effect is dependent on the molecular mass and concentration of the added PEG. The theory of the preferential exclusion of PEG was used to explain the mechanism. The presence of PEG was hypothesized to elicit an energetically favorable sharing of the cosolvent exclusion shells surrounding the proteins and chromatography media, and hence to elevate partition coefficients[14]. Although addition of PEG also increase viscosity, with the attendant affects of reducing flow-rate and dynamic binding capacity, while increasing eluted peak width, addition of PEG may have useful preparative application among coeluting proteins of significantly similar size, i.e. PEG can produce potentially useful compound selectivity.

The above mentioned chromatographic experiments with PEG were carried out using commercially purified reagent-grade proteins rather than purifying them from a natural mixture such as cell homogenate. Besides, the activity of the proteins after chromatography was not measured. In order to investigate the practical usefulness of PEG in real separation, purification of recombinant human tumor necrosis factor-α (TNF-α) from *E coli* was investigated as a model system. Figure 2 and 3 demonstrated the comparative results of ion exchange purification of TNF-α without PEG (PEG=0) and with PEG (PEG 200, 600, and 4000) in the feed. When there was no PEG present, the recovery could only reach about 65% as shown in Figure 2 (PEG=0), and the purification factor was about 7 as shown in Figure 3. With addition of PEG in the feed, both the recovery and the purification factor were changed. The optimum was shown at 1% for the three PEGs with PEG200 the best . The recovery even surprisingly showed to more than 100% , indicating that part of the denatured product in the feed might be renatured. The purification factor was doubled to 14.

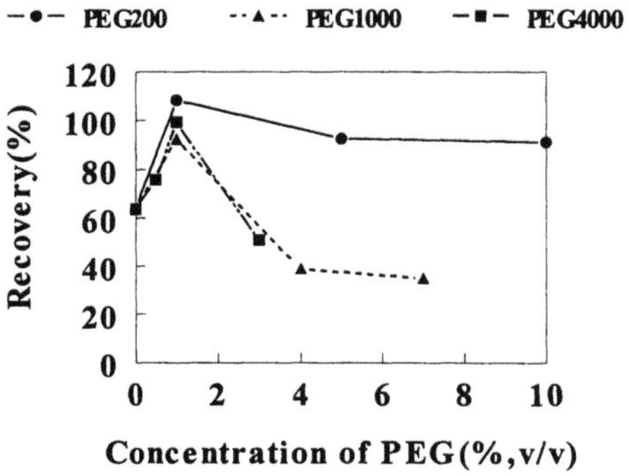

Figure 2 Variation of recovery during ion exchange chromatography of TNF-α

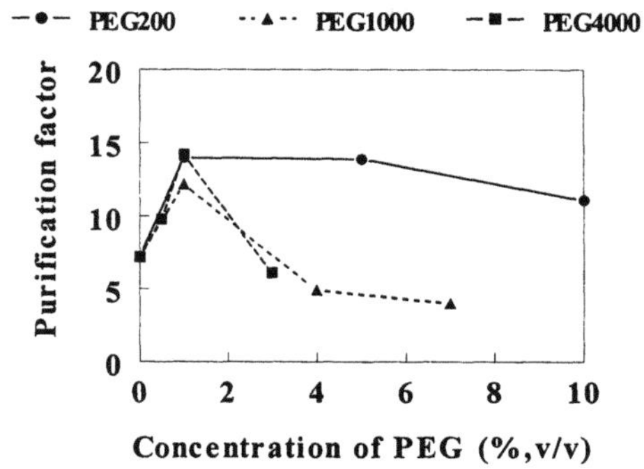

Figure 3 Variation of purification factor for ion exchange chromatography of TNF-α

The strategies proposed above depends on different proteins. It is important to know the structure and property of the target product such as its thermal stability, stable pH range, etc.[15], and to design the bioseparation route accordingly. For example, if the target protein is thermal stable, elevated temperature may not denature it while other impurities may precipitate, which is a simple and effective way of initial purification.

REFERENCES

1. C.J. Gray, In: Recovery Process for Biological Materials, J.F. Kennedy and J.M.S Cabral (eds), John Wiley & Sons , New York, 1993.
2. A.S. Chandavarkar, PhD Thesis, Massachusetts Institute of Technology, 1990.
3. S. Li, C. Schoneich and R. T. Borchardt, Biotechnol. Bioeng., 48 (1995) 490.
4. M. Kaufmann, J. Chromatogr. B, 699 (1997) 347.
5. P.R. Foster, P. Dunnil and M.D. Lilly, Biochim. Biophys. Acta, 317 (1973) 505.
6. A. McPherson Jr., J. Biol. Chem., 251 (1976) 6300.
7. B.A. Andrews and J.A. Asenjo, In: Protein Purification Methods-A practical Approach. E.L.V. Harris, and S. Angal, (eds.), IRL Press, Oxford, 1989.
8. P.A. Albertsson, Partition of Cell Particles and Macromolecules, 3rd Edition, John Wiley and Sons, New York, 1986.
9. S.N. Timasheff, In: Stability of Protein Pharmaceuticals (Part B), T.J. Ahern and M.C. Manning (eds.), Plenum Press, New York, 1990.
10. J.L. Cleland, S.E. Builder, J.R. Swartz, M. Winkler, J.Y. Chang, and D.I.C. Wang, Bio/Technol., 10 (1992) 1013.
11. T. Arakawa and S.N. Timasheff, Biochem., 24 (1985) 6756.
12. S-C.B. Yan, D.N. Tuason, V.B. Tuasonand W.H. Frey II, Anal. Biochem., 138 (1984) 137.
13. C.L.De. Ligny, W.J. Gelsema and A.M.P. Roozen, J. Chromatogr., 294 (1984) 223 .
14. P. Gagnon , B. Godfrey and D. Ladd, J. Chromatogr. A ,743 (1996) 51.
15. N.P. Pace, Trends in Biotechnol., 8 (1990) 93.

Bioseparation Engineering
I. Endo, T. Nagamune, S. Katoh and T. Yonemoto (Editors)

Bioseparation of Natural Products

Craig Keim and Michael R. Ladisch

Laboratory of Renewable Resources Engineering and Department of Agricultural and Biological Engineering, Purdue University, West Lafayette, IN 47907

Bioseparations engineering is the application of fundamental engineering and biological principles to the design of adsorbents, equipment and processes for the separation of biological molecules. Research and development of bioseparation processes combines the disciplines of engineering, life sciences, chemistry and medicine in order to match the molecular properties of biomolecules with the most appropriate techniques for their large scale purification. Knowledge of the controlling mechanisms of individual separation steps, once known, enables fractionation methods to be selected. These steps are then combined to give processing sequences that result in product purification at an acceptable cost and in a reasonable period of time.

Certain natural products derived from plant, animal, and marine tissues, as well as those harvested from the in-vitro cultivation of microorganisms have therapeutic, nutritional, or biochemical value. The biological extracts derived from plant and animal tissue may consist of mixtures of proteins, polysaccharides, or secondary metabolites that have a relatively low molecular weight. Purification is therefore needed to obtain products that are suitable for human or animal consumption, or for use as specialty biochemicals. Consequently, separation methods are needed to efficiently recover and purify products from natural materials.

This paper addresses the special characteristics of biochemical mixtures derived from natural sources in the context of their purification by chromatographic separations. The role of bioseparations engineering in designing systems to purify these products using environmentally compatible methods is discussed. Two case studies are presented that illustrate the principles and benefits of the naturally derived, renewable materials, cellulose and starch, as separations media for purification of natural products by adsorptive and chromatographic methods.

Introduction

The purification of proteins and other bioproducts is a critical and expensive part of most biotechnology based manufacturing processes, and may account for 50% or more of production costs (1). While overall production costs have been considered to be secondary to being the first to market, this perspective is changing as the price - and value - of new bioproducts is decreasing. When the volume of the products is small and the price is high, being the first to market, together with attaining high product quality (in terms of purity, activity, dependability, or flexibility) are the major competitive advantages (1-3). Bioseparations are important in assuring product quality, but manufacturing cost is secondary for these types of products.

As the scale of production of new bioproducts continues to grow from kilograms to tons, the need for cost-effective purification schemes is also increasing in importance. High volume products range from serum proteins produced by recombinant organisms to organic acids, enzymes, and food additives obtained from large scale fermentations or enzyme transformations.

One of the major technical challenges in the production of pharmaceuticals is the "development of high-resolution protein purification technologies that are relatively inexpensive, are easily scaled-up and have minimal waste-disposal requirements (1)." Separation processes for bioprocessing of renewable resources and agricultural products will benefit from development of "more efficient separations for recovering fermentation products, sugars, and dissolved materials from water," and in particular, lowering the cost of separating water from the product in the fermentation broth (1). These challenges can be addressed through chromatographic, membrane, and adsorptive separations.

The prospects of chromatographic separations continue to grow, particularly as separations of chiral compounds, protein pharmaceuticals, and value added bioproducts from agriculture become an important determinant of product quality. Historically, chromatography has been a relatively slow and expensive technology. The challenges lie in developing new adsorbents and chromatographic stationary phases that maximize mass transfer area per unit column volume and minimize mass transfer resistance. These stationary phases must also have robust hydraulic and chemical operating characteristics.

Biomaterials As Separating Agents

The removal of water from ethanol and from other types of vapors or gases uses a biomaterial and renewable resource, corn, as the adsorbent (4, 5). This starch-based adsorbent adsorbs water much more rapidly than ethanol (4) and enables a dry product to be obtained from hydrous vapor from a distillation (stripping) column (6). The ground corn in a packed bed is initially dry and pretreated to a temperature of 80 to 90°C. The alcohol/water vapors are passed over the bed. The concentration and temperature profiles move in close proximity to one another as the water adsorbs. Breakthrough of the water concentration profile coincides with a sharp increase in the temperature - and hence temperature provides a convenient measure for monitoring the adsorption process. This process is now used in place of azeotropic distillation to dry approximately 750 million gallons of fuel ethanol, annually, in the US.

Analogues of corn adsorbents, synthesized from starch, have recently been developed, characterized, and tested as a drop-in replacement for molecular sieves in a laboratory-scale pressure swing dryer to dry air to between -70 to -80°C dew points (7-9). Development of these materials combined with engineering evaluation and modeling of transport properties is a cooperative effort with researchers in the Departments of Material Engineering and Mechanical Engineering at Purdue University. Continuing fundamental research on equilibrium and transport properties for these materials will help to facilitate design and scale-up of novel polysaccharide adsorbents for new applications. These are envisioned to include a range of applications from drying of industrial gases to desiccant-based air conditioners where biodegradable adsorbents would supplant freons in residential and commercial applications.

Affinity ligands are another example of a biomaterial that serves as a separating agent. These ligands are proteins derived from microorganisms and mammalian cells. Various types of affinity ligands have been demonstrated to be effective in purifying therapeutic proteins at the laboratory scale. Several manufacturing processes in the biotechnology industry use them as part of protein purification sequences (2). The ligands, once identified and obtained in a large enough quantity for process applications, are immobilized or chemically attached to the stationary phase. A solution containing the protein bioproduct is then passed over a column of the immobilized antibody at conditions that facilitate selective binding of the protein to the antibody. The impurities (which do not bind) are washed away. A change in the mobile phase then causes the protein to dissociate from the immobilized antibody so that it elutes in a purified form. The column is then re-equilibrated with the starting buffer, and the process is repeated. While this method has seen some industrial use, its growth as a process separation tool requires development of techniques that can produce large quantities of the ligands (i.e. on a kilogram scale) at a reasonable price. Robustness and validation that small amounts of the ligand do not leak into the product are also important for this type of separation to gain acceptance on the process scale.

Regulation of Manufacturing Processes for Biologics and Drugs

The production of biosynthetic human insulin by microbial fermentation requires 31 major processing steps of which 27 are associated with product recovery and purification (11). After the insulin is produced in an *E. coli* fermentation the cells are lysed so that the inclusion bodies are released. The inclusion bodies are then dissolved, and the protein is refolded into a conformation that will eventually lead to an active molecule. Reagents used in these steps are later removed when insulin is purified by a series of ion exchange, reverse phase, and size exclusion chromatography steps (2, 12, 13). The purification of insulin not only illustrates the many steps involved, but also that chromatography steps, which are based on 50 to 1000 L of adsorbent, are large in the context of biotechnology manufacturing but modest by chemical industry standards.

Tissue plasminogen activator (t-PA) is a proteolytic enzyme derived from a recombinant cell line, which is capable of thrombolysis (dissolving of blood clots) during a heart attack (10, 14). Recombinant technology provides the only practical means of producing this pharmaceutical since one dose is about 100 mg. A volume of 50,000 L of blood (containing 2 to 5 ng/L of t-PA) would otherwise be needed to produce one dose. Cell lines consisting of transformed (genetically engineered) Chinese Hamster Ovary cells (abbreviated CHO) selected for high levels of t-PA expression are used to generate this protein. In this case, a bacterium such as *E. coli* cannot be used because the t-PA must be properly glycosylated (10, 15).

The purification of recombinant t-PA may include the steps of: (i) cell removal by sterile filtration; (ii) protein purification accompanied by DNA and virus removal; and (iii) final purification by ion exchange and size exclusion chromatography. The possibility that DNA from an immortal cell line such as CHO cells could cause oncogenic (gene altering) events was addressed during development of the purification sequence (10). While the DNA by itself was shown to be inactive *in vivo*, when injected into rodents, the removal of DNA to less than 10 picogram/dose (1 picogram = 10^{-12} gram) needed to be achieved as part of the manufacturing process (10).

Small Molecules Separation and Purification

Small molecules are derived through fermentation, biochemical modification of fermentation products or chemical synthesis. These include antibiotics, vitamins, nucleosides, alcohols, and organic acids. These are particularly amenable to purification using reversed phase chromatography (16, 17). Process scale adsorption and chromatography processes for these molecules are likely to use silica gels, polymeric adsorbents, or derivatized polymeric adsorbents (i.e., ion exchange resins).

Small molecules, unlike proteins, are unlikely to change in conformation during purification, and can be processed using organic solvents and different forms of reversed phase chromatography. Reversed phase chromatography utilizes increasing concentrations of aqueous alcohols, acetonitrile, acetone, ethyl acetate, or hexane to separate molecules of different polarities (18). Consequently, the pairing of an appropriate mobile phase composition with a given adsorbent allows the separation of molecules that differ only slightly in their polarity. Examples of products which utilize reversed phase chromatography as part of their purification protocols are Salmon calcitonin (19), cefonicid (20), and diastereomer precursors for the insect sex pheromone from Lamantria dispar (21). Calcitonin is a 32 residue peptide used for treatment of post-menopausal osteoporosis, hypercalcemia, and Paget's disease. Cefonicid is an intermediate in the production of β-lactam antibiotics, while the pheromone is used to control a pest that attacks oak trees.

Improvements in bioseparation techniques will help to improve yields of small molecules and lower molecular weight products. However, the application of recombinant technology will also be important in removing bottlenecks in synthetic pathways leading to the bioproducts, as recently discussed in literature surveys on Cephalosporin C (22, 23).

Elucidation of the biosynthetic pathway for production of Cephalosporin C in *Cephalosporium acremonium* resulted in identification of a bottleneck associated with the enzyme which converts penicillin N to a cephalosporin C precursor. Queener, Skatrud and his colleagues introduced extra copies of the gene responsible for synthesizing the enzyme into *C. acremonium* using recombinant methods. This resulted in about 20 to 40% enhanced Cephalosporin C production on a laboratory scale, and 15% improvement on the pilot plant scale.

Hence, recombinant technology increased product yield and concentration by enabling insertion of an extra copy of the gene for a rate-limiting enzyme. This type of improvement could give impressive increases in productivity, but only if product recovery and efficiency of the bioseparations steps are improved to yield higher recoveries and throughputs during downstream processing.

Summary

The field of biochemical separations is growing as the number and volume of biotechnology products increases, with the cost of their production, as well as their purity, becoming important issues. The process scale purification of these products can make up 50% or more of their manufacturing cost with various forms of chromatographic and membrane separations being major parts of fractionation sequences used for purifying these products. The current status of chromatographic and adsorptive separation techniques shows that development of separations media and adsorbents which are resistant to fouling, have minimal mass transfer resistances, and possess long term stability are important factors in expanding their use in the biotechnology industries. Fundamental modeling of the basic separation and hydraulic mechanisms will be important for developing new applications of existing materials, as well as moving new separations media from the laboratory to the plant. The engineering of separation processes of biotechnology products requires a fundamental understanding of the properties of the biological materials being processed as well as the principles of bioseparations unit operations.

Acknowledgments

The material in this work was supported by Purdue University Agricultural Research Programs Office and the Laboratory of Renewable Resources Engineering. I thank Dr. Joe Weil and Kyle Beery for helpful comments, thoughtful analysis, and stimulating discussions during preparation of this manuscript.

References

1. Committee on Bioprocess Engineering, National Research Council, *Putting Biotechnology to Work: Bioprocess Engineering*, National Academy of Sciences, Washington, DC, 2-22 (1992).

2. S. M. Wheelwright, "Protein Purification: Design and Scale up of Downstream Processing," Hanser Publishers, Munich, 1-9, 61, 213-217 (1991).

3. C. A. Bisbee, "Current Perspectives on Manufacturing and Scale-up of Biopharmaceuticals," *GEN*, *13*(14), 8-9 (1993).

4. Lee, J. Y., P. Westgate, and M. R. Ladisch, "Water and Ethanol Sorption Phenomena on Starch," *AIChE J.*, *8*(37), 1187-1195 (1991).

5. Westgate, P. J., and M. R. Ladisch, "Air Drying Using Corn Grits as the Sorbent in a Pressure Swing Adsorber," *AIChE J.*, *39*(4), 720-723 (1993).

6. M. R. Ladisch, M. Voloch, J. Hong, P. Bienkowski, and G. T. Tsao, "Cornmeal Adsorber for Dehydrating Ethanol Vapors," *Ind. Eng. Chem. Des. Dev.*, *23*, 437-443 (1984).

7. Anderson, L., M. Gulati, P. Westgate, E. Kvam, K. Bowman, and M. R. Ladisch, "Synthesis and Optimization of a New Starch Based Adsorbent for Dehumidification of Air in a Pressure Swing Drier," *Ind. & Eng. Chem. Res.*, *35*, 1180-1187 (1996).

8. Westgate, P., and M. R. Ladisch, "Sorption of Organics and Water on Starch," *Ind. Eng. Chem. Res.*, *32*(8), 1676-1680 (1993).

9. Westgate, P. J., J. Y. Lee, and M. R. Ladisch, "Modeling of Equilibrium Sorption of Water Vapor on Starch Materials," *Transactions ASAE*, *35*(1), 213-219 (1992).

10. S. E. Builder, R. van Reis, N. Paoni, and J. Ogez, "Process Development and Regulatory Approval of Tissue-Type Plasminogen Activator," in Proceedings of the 8th International Biotechnology Symposium, Paris (July 17-22, 1989).

11. W. F. Prouty, "Production-Scale Purification Processes," in *Drug Biotechnology Regulation*, *13*, Y-Y. H. Chien and J. L. Gueriguian, ed. Marcel Dekker, NY, 221-262 (1991).

12. D. P. Petrides, J. Calandranis, C. L. Cooney, "Bioprocess Optimization Via CAPD and Simulation for Product Commercialization," *GEN, 16*(16), 24, 28 (1996).

13. M. R. Ladisch and K. L. Kohlmann, "Recombinant Human Insulin," *Biotechnol. Prog., 8*(6), 469-478 (1992).

14. S. E. Builder and E. Grossbard, "Laboratory and Clinical Experience with Recombinant Plasminogen Activator," in *Transfusion Medicine, Recent Technological Advances,* 303-313 (1986).

15. J. D. Watson, M. Gilman, J. Witkowski, and M. Zoller, *Recombinant DNA*, Second Edition, W. H. Freeman and Company, NY, 458-460 (1992).

16. H. Takayanagi, J. Fukuda, and E. Miyata, "Non-ionic Adsorbents in Separation Processes," in *Downstream Processing of Natural Products, A Practical Handbook*, M. Verrall, ed., J. Wiley andSons, Chichester, 159-178 (1996).

17. F. X. Pollio and R. Runin, "The Use of Macroreticular Ion Exchange Resins of the Fractionation and Purification of Enzymes and Related Proteins," *Chem. Eng. Symp. Ser., 67*(108), 66-74 (1971).

18. P. C. Sedek, P. W. Carr, R. M. Doherty, M. J. Kamlet, R. W. Taft, and M. H. Abraham, "Study of Retention Processes in Reversed-Phase High-Performance Liquid Chromatography by the Use of the Solvatochromic Comparison Method," *Anal. Chem., 57*, 2971-2978 (1985).

19. E. Flanigan, (Rhône Poulenc Rorer), "High Performance Liquid Chromatography in the Production and Quality Control of Salmon Calcitonin," in Purdue University Workshop on Chromatographic Separations and Scale-up, 207 (1991).

20. A. M. Cantwell, R. Calderone, and M. Sienko, "Process Scale-up of β-Lactam Antibiotic Purification by High Performance Liquid Chromatography," *J. Chromatogr., 316*, 133-149 (1984).

21. G. Pierri, P. Piccardi, G. Muratori, L. Cavalo, "Scale-up for Preparative Liquid Chromatography of Fine Chemicals," *La Chimica E L'Industria, 65*(5), 331-336 (1983).

22. J. Weil, J. Miramonti, and M. R. Ladisch, "Cephalosporin C: Mode of Action and Biosynthetic Pathway," *Enz. Microb. Technol., 17*(1), 85-87 (1995).

23. J. Weil, J. Miramonti, and M. R. Ladisch, "Biosynthesis of Cephalosporin C: Regulation and Recombinant Technology," *Enz. Microb. Technol., 17*(1), 88-90 (1995).

Bioseparation Engineering
I. Endo, T. Nagamune, S. Katoh and T. Yonemoto (Editors)

On-line recovery of large molecules from mixture solution using semi-continuous size exclusion chromatography

Y.-M. Kim, W.-J. Chang and Y.-M. Koo

Department of Biological Engineering, Inha University, Inchon 402-751, Korea

The recovery of Blue Dextran from the mixture solution with vitamin B_{12} was carried out using the reciprocating size exclusion chromatography (RSEC). The separation performances in RSEC were compared with those in the conventional size exclusion chromatography (SEC) with recycle. The recovery yields of Blue Dextran in RSEC and SEC with recycle after five cylces were 64% and 51 %, respectively.

1. INTRODUCTION

Size exclusion chromatography separates solutes of different size, based upon the size exclusion effect of porous gels packed in a column. SEC has been employed commercially for purification of plasma proteins and as one step in purification of many other proteins, and analytically for protein separation and for analyzing polymer molecular weight distributions. Elutions in SEC are carried out mostly in peak mode, where solutes of different molecular size in a pulse of feed are separated from each other and exit from the column as separate peaks. The peak mode elution is a common practice in analytical chromatography. Operating methods are discussed in a various articles and books [1].

A modified operation of size exclusion chromatography, RSEC, was developed to recover large molecules on-line from the mixture solution [2]. On-line recovery of large molecules from the mixture is an unusual trial, comparing to the routine practice of filtration where small molecules are isolated from the mixture. RSEC is operated semi-continuously, based upon the elution in frontal mode, where solutes of different size in a step feed proceed along the column, forming fronts of their own. The frontal mode operation is considered to give higher separation capacity than the peak mode in the preparative chromatography. In this study, the separation performances in RSEC were compared with those in SEC with recycle in which the isolated small solute from the pulse input of mixture solution was recycled to the feed mixture

2. MATERIALS AND METHODS

A commercial polyacrylamide gel, Bio-Gel P-10 (Bio-Lad, fractionation range: 1,500-20,000), was packed in a water-jacketed column (Pharmacia SR 10/50). The packed gel layer was compressed from both sides of the column using plungers, with the degree of compression of 0.82 [3]. The final length of the compressed gel was 24.5 cm. The temperature of the column was kept at 25 °C using a constant-temperature water bath (Cole-Parmer) and a water jacket. Elution behavior in a SEC column (1 cm (ID) x 24.5 cm), was tested using Blue Dextran (Pharmacia, MW: 2,000,000) and vitamin B_{12} (BDH Laboratory Supplies, MW: 1355.38) as standard materials for large and small molecules, respectively. The feed concentrations of Blue Dextran and vitamin B_{12} were 0.2 g/l and 0.02 g/l, respectively. The initial volumes of the feed reservoir were 40 ml and 30 ml in RSEC and SEC with recycle, respectively. The eluent flow rate was 0.42 ml/min in both directions. The reciprocating operation was carried out using a high-pressure pump (Eldex) and a multi-way valve (Cole-Parmer) on each side of the gel-packed column, controlled by a multiport programmable timer (ChronTrol). Concentrations of the two solutes were calculated using the binomial equations from the optical absorbances (Spectronic) at 615 nm and 361 nm. These are the wavelengths of the maximum absorption of the two solutes.

In RSEC (Fig. 1, (A)), the large molecules (dextran polymers) were isolated from the mixture by repeating cycles of feeding mixture solution. The large molecules were isolated into large solute tank during the forward flow period in frontal mode and the following slow-moving portion of unseparated mixture solution was returned to the reservoir by backward flow. The solvent eluted before the large molecules during the first half cycle was gathered in solvent reservoir, and reused as eluent for the backward flow during the second half cycle.

In SEC with recycle (Fig. 2, (B)), the feed mixture was fed to the column as a pulse, followed by an eluent. The isolated band of the small molecules (vitamin B_{12}), following the band of pure large molecules, was recycled to the feed reservoir. The durations of the feed pulse and the eluent in each cycle were scheduled so that the band of pure small molecules

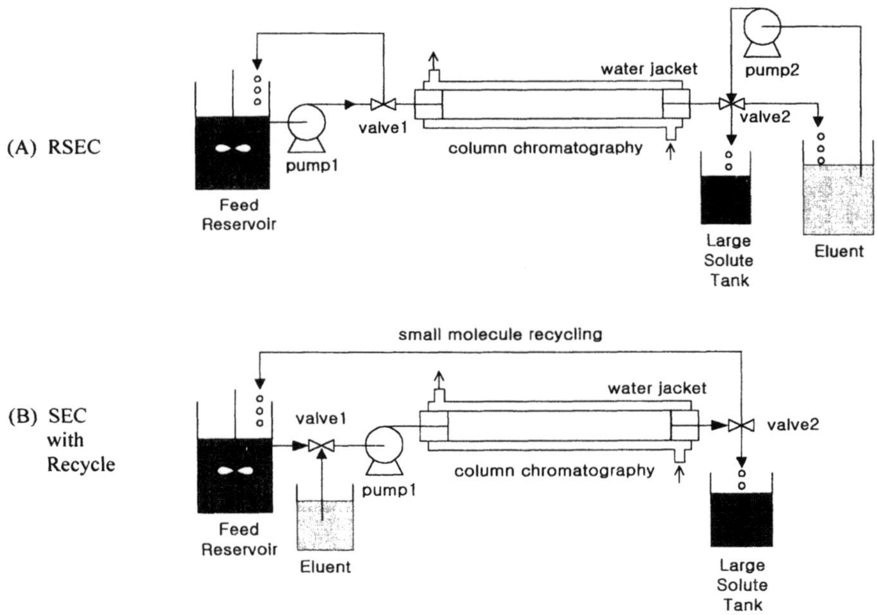

Figure 1. Schematic drawings of modified SEC.

was touched at the base line by the following band of pure large molecules from the next cycle at the exit of the column.

3. RESULTS AND DISCUSSION

In frontal mode operation of RSEC, solutes of different size in a step feed proceed along the column, forming fronts of their own (Fig. 2). From this elution curves, the retention volumes of Blue Dextran and vitamin B_{12} at 25°C were calculated to be 6.9 ml and 17.6 ml, respectively. Initially, the packed column was filled with pure solvent, and the reservoir contained 40 ml of the feed mixture solution. In the first half cycle of pumping solution from the reservoir to the column (forward flow), Blue Dextran was recovered as a second fraction of 7.4 ml between 12.5 min and 30 min, following the first fraction of pure solvent of 5.3ml (Fig. 2). One reciprocating cycle was completed by pumping 13.6 ml of pure solvent back to the reservoir during the second half cycle of 42.4 min (backward flow). The concentration changes of solutes in the reservoir and the recovery tank were shown in Fig. 3. The concentrations of Blue Dextran and vitamin B_{12} in the figure were presented as percent compared with the initial concentrations. With the repeating cycles, the amount of Blue Dextran in the reservoir decreased, while the amount of vitamin B_{12} remained constant, as no significant amount of vitamin B_{12} was found in the recovery tank. This relatively small loss of vitamin B_{12} was also confirmed by calculation from the vitamin B_{12} concentration in the reservoir and the volume of the reservoir. Note that the volume of the backward flow to the reservoir in the second half cycle was more than that of the forward flow in the first half cycle by 1 ml per cycle, resulting in the dilution of solutes in the reservoir. This difference in volumes protected the recovered solution from being contaminated by the vitamin B_{12} which diffused from the moving front of the mixture in the column. A recovery rate of 64 % was observed with Blue Dextran after the 5th cycle. The amount of the recovered solute can be calculated based upon a simple mass balance in the system.

In SEC with recycle, a pulse feed of the mixture solution was fed to the same column as in RSEC, followed by eluent to push the feed solution. The durations of the feed and the eluent pumping were 12 min and 40.5 min, respectively, to have the trains of alternating pure

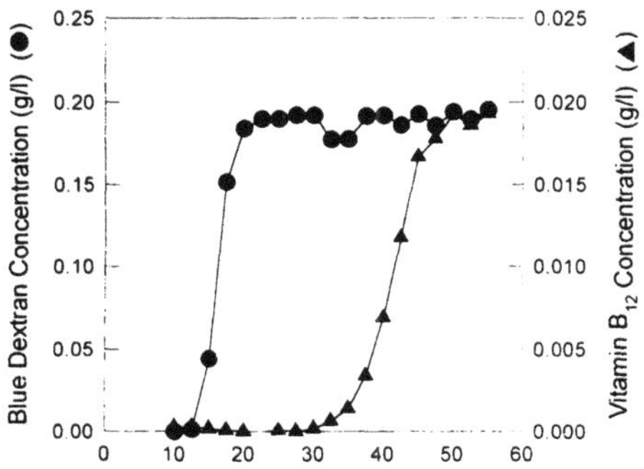

Figure 2. Elution curves of Blue Dextran and vitamin B_{12} for RSEC.

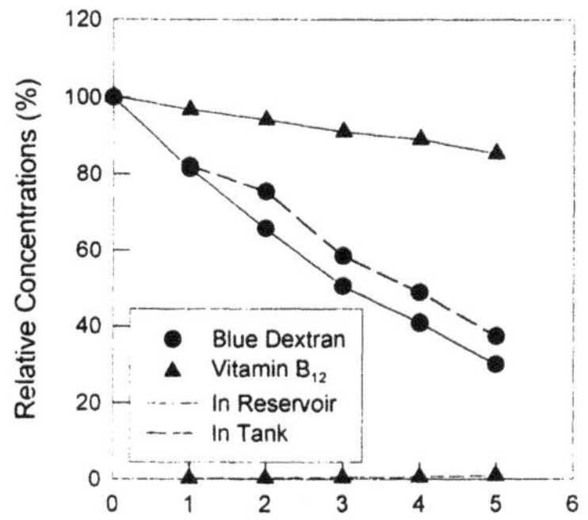

Figure 3. Separation performances in RSEC.

bands of large and small solute exit the column. A train of separated bands of pure Blue Dextran and vitamin B_{12} was shown in Fig. 4. Only the band of Blue Dextran was recovered in the tank, while the following band of vitamin B_{12} for 30 min was recycled to the feed reservoir. The concentration changes of the solutes in the feed reservoir and the recovery tank in SEC with recycle were shown in Fig. 5. The concentration of Blue Dextran in the feed reservoir decreased faster than that of vitamin B_{12} with repeating cycles. The concentration of vitamin B_{12} in the reservoir decreased as cycles repeated, because the solution volume in the reservoir increased by 7.56 ml per cycle. The amount of vitamin B_{12} in the reservoir was considered to be constant as a negligible amount of vitamin B_{12} was observed in the recovered solution. The recovery rate of Blue Dextran was calculated from the Blue Dextran concentration in the reservoir to be 51 % after five cycles.

The recovery rate in RSEC was higher that that in SEC with recycle by 13 %. The main part of this difference is considered to be caused by the thermodynamic inefficiency in SEC with recycle, in the sense that the purified vitamin B_{12} was recycled and remixed with the feed solution in the reservoir. The operating conditions, mainly the timing schedules, were confirmed to be near to the optimal values, by repeating elution experiments in a way of trial and error. During the cyclic operations, the mass balance in the system, including the reservoir, gel column, and recovered solutions, was met within the error range of 5%. At the present time, mathematical simulations, based upon the local equilibrium model, are being carried out to compare the productivities of various operation types of SEC, such as RSEC, SEC with recycle, and SEC with column switching, by the current authors.

REFERENCES

1. W.W. Yau, J.J. Kirkland and D.D. Bly, Modern Size Exclusion Liquid Chromatography, Wiley, New York, 1979.
2. W.-J. Chang and Y.-M. Koo, Biotech. Tech., 13 (1999) 211.
3. Y.-M. Koo and P.C. Wankat, Ind. & Eng. Chem. Fund., 24 (1985) 108.

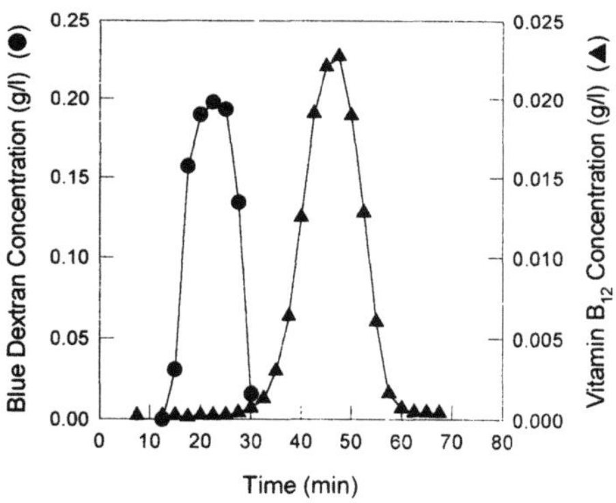

Figure 4. Elution curves of Blue Dextran and vitamin B_{12} for SEC with recycle.

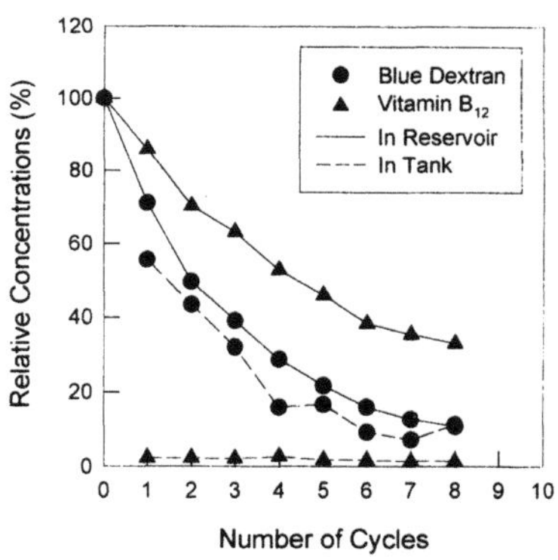

Figure 5. Separation performances in SEC with recycle.

Bioseparation Engineering
I. Endo, T. Nagamune, S. Katoh and T. Yonemoto (Editors)

Dye Adsorption by Activated Carbon in Centrifugal Field

Chia-Chang Lin and Hwai-Shen Liu*

Department of Chemical Engineering, National Taiwan University,
Taipei, Taiwan, ROC
*E-mail: hsliu@ccms.ntu.edu.tw

The adsorption of two dyes, namely, Basic Yellow 2 and Reactive Orange 16 on activated carbon from aqueous solutions under centrifugal field was studied. The results indicated that the centrifugal force could enhance the adsorption rate for both dyes. The data also showed that the centrifugal force could increase the adsorption rate with a higher degree for Reactive Orange 16 due to a lower mass transfer resistance. Consequently, the mass transfer for dye adsorption could be improved using the centrifugal force.

1. INTRODUCTION

In recent years, mass transfer intensification by vapor-liquid contact in a centrifugal field attracts some attention. That is achieved by rotating a torus-shaped packing element. This concept was first proposed by Ramshaw and Mallinson (1) in 1981 and named by "Higee" (high gravity). The liquid is subjected to a high acceleration of at least 300 m/s^2, tuned by rotation speed, as it flows through the packing element. As a result, the tendency of flooding compared with that in a conventional packed bed could be reduced. Thus, higher gas and liquid flow rates could be used and the specific area of the packing could be increased. Moreover, the liquid film would become thinner and mass transfer may also be enhanced by 10-100 fold. Therefore, the physical size of the equipment would be greatly reduced in comparison with a conventional packed bed and, thus, lower capital and operating cost would be reduced. A few literatures have been published concerning about the applications of centrifugal field in gas-liquid systems such as distillation, absorption, stripping and deaeration (2-10). However, the application of centrifugal field to a liquid-solid system such as adsorption is not well-known to date. In order to investigate the characteristics concerning the effect of centrifugal force on adsorption, the bed adsorption with recycle system is adopted. The adsorption process chosen is a dye adsorption on activated carbon from an aqueous solution.

2. EXPERIMENT

The main objective of this work is to examine if the centrifugal force could affect the performance of adsorption. The dyestuffs, Basic Yellow 2 (supplied by Acros) and Reactive Orange 16 (supplied by Aldrich), were used as the adsorbate in this work. The activated carbons used as the adsorbent in this work were cylindrical activated carbon (0.1-1 cm length;

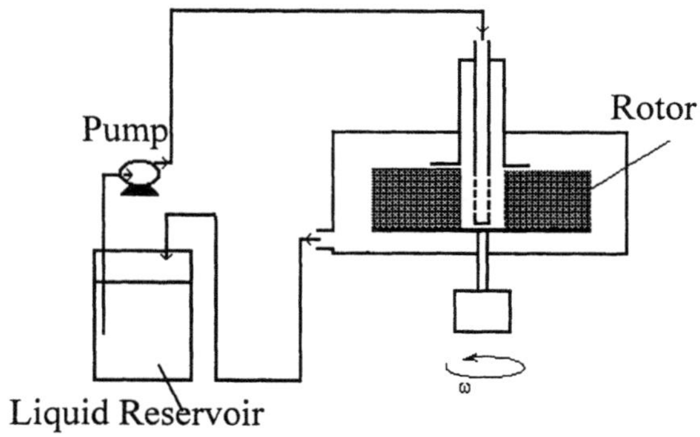

Figure 1. Centrifugal adsorption bed with recirculation

0.4 cm diameter) (supplied by HOTAI). The activated carbon was dried at 110 °C for more than 24 hr before experiments.

Figure 1 shows the simplified schematic diagram of adsorption bed with recirculation. The centrifugal adsorption bed comprises a rotor and a stationary housing. Liquid flows through the adsorption bed outward from the inner surface of the rotor due to the centrifugal force. For visual observation, the rotor and housing are made of transparent acrylic. The housing has an internal diameter of 12 cm. The bed has an inner radius of 2 cm, an outer radius of 4 cm, and an axial height of 2 cm. The length of liquid path is 2 cm. The total bed volume is 75.4 cm^3. The activated carbon with known weight is packed randomly within the bed. With the variable speed motor, the bed can be operated from 400 to 2500 rpm, which provides 5 to 210 gravitational force based on the arithmetic mean radius.

For a typical experiment, an aqueous dye solution in the reservoir (liquid volume = 2.5 L) was introduced to the top of the equipment and flowed through the activated carbon bed, and then expelled from the bottom of the equipment, recycled back to the reservoir. In operation, the liquid flowed over activated carbon as film. Thus, the radial velocity of the liquid within the rotor would depend on the rotation speed. In all runs, samples were taken from the reservoir and analyzed with a spectrophotometer (SPECTRONIC 20 GENESYS) at wave length of 476 nm for Basic Yellow 2 and 535 nm for Reactive Orange 16. Investigated process parameters included the centrifugal force and the initial dye concentration. The weight of cylindrical activated carbon packed within the bed was 36 g, the recirculation flow rate was 1532 mL/min and the temperature was 31°C.

3. RESULTS AND DISCUSSION

Experimental results for the adsorption of two dyes on cylindrical activated carbon with different initial dye concentrations (205, 165, 125 and 85 mg/L) are shown in Figure 2. Each sub-graph illustrates a plot of the adsorbed fraction against time with various rotor speed (0, 400, 1000, 1600 rpm). These rotor speeds provides a centrifugal acceleration variation from 0 to 842 m/s^2 based on the average mean radius. The data indicated that the centrifugal force indeed provides the improvement in the performance of dye adsorption by activated carbon

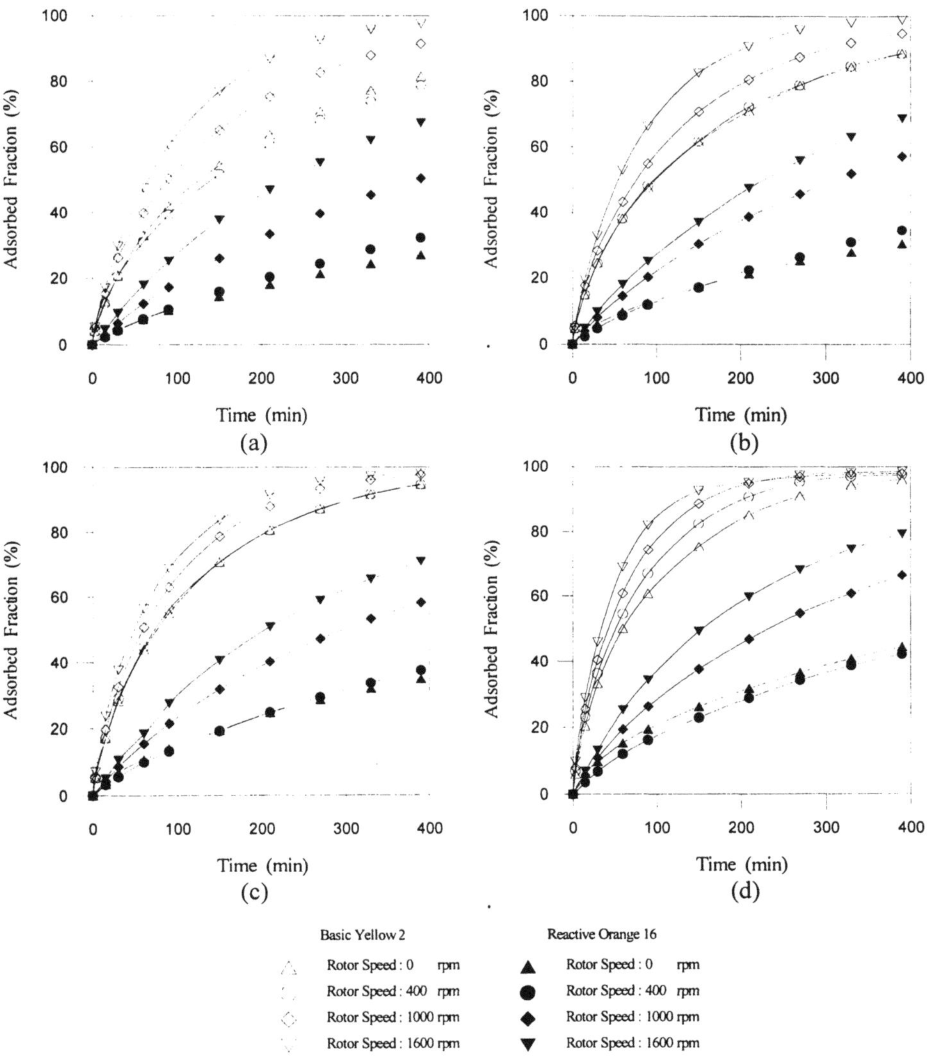

Figure 2. Comparison of dye adsorption for various rotor speeds(initial dye concentration = (a) 205 mg/L (b) 165 mg/L (c) 125 mg/L (d) 85 mg/L)

for all initial concentrations. That is, the rate of the dye adsorption could be enhanced using the centrifugal force. This may be due to the reduced resistance of mass transfer provided by the centrifugal force. It is also found that the enhancement in the adsorbed fraction due to the centrifugal force depends on the dye initial concentration. The increase of the adsorbed fraction by the centrifugal force becomes more obvious for high initial concentration. This is probably due to that the centrifugal force may decrease the collisions among dye molecules at high concentration which otherwise would increase the resistance of mass transfer.

As shown in four sub-graphs of Figure 2, the enhancement of the adsorbed fraction due to the centrifugal force for reactive dye is larger than for basic dye. This phenomenon may be due to the size difference of dye molecules. The reactive dye of larger size is difficult to move onto the surface of the activated carbon without the centrifugal force. However, the centrifugal force provides a lower resistance for larger dye molecules. As a result, the adsorbed fraction of reactive dye is increased with a greater degree under centrifugal field.

4. CONCLUSIONS

The experimental results of dye adsorption by activated carbon under centrifugal field were investigated. The data indicated that the rate of adsorption can be controlled by the degree of centrifugal force for both dyes. The results also showed that the enhancement in the performance of the adsorption due to the centrifugal force is larger for reactive dye than for basic dye because of the reduced mass transfer resistance. Therefore, the dye adsorption could be improved by the centrifugal force.

REFERENCES

1. C. Ramshaw and R. H. Mallinson, Mass Transfer Process. US Patent No. 4 283 255 (1981).
2. M. Keyvani and N. C. Gardner, Chem. Eng. Prog. 85 (1989) 48.
3. S. Munjal, M. P. Dudukovic and P. A. Ramachandran, Chem. Eng. Sci. 44 (1989) 2245.
4. M. P. Kumar and D. P. Rao, Ind. Eng. Chem. Res. 29 (1990) 917.
5. S. P. Singh, J. H. Wilson, R. M. Counce, J. F. Villiers-Fisher, H. L. Jennings, A. J. Lucero, G. D. Reed, R. A. Ashworth and M. G. Elliott, Ind. Eng. Chem. Res. 31 (1992) 574.
6. A. Basic and M. P. Dudukovic, AIChE J. 41 (1995) 301.
7. H. S. Liu, C. C. Lin, S. C. Wu and H. W. Hsu, Ind. Eng. Chem. Res. 35 (1996) 3590.
8. T. Kelleher and J. R. Fair, Ind. Eng. Chem. Res. 35 (1996) 4646.
9. F. Guo, C. Zheng, K. Guo, Y. Feng and N. C. Gardner, Chem. Eng. Sci. 21/22 (1997) 3853.
10. J. Peel, C. R. Howarth and C. Ramshaw, Trans IChemE. PartA. 76 (1998) 585.

Bioseparation Engineering
I. Endo, T. Nagamune, S. Katoh and T. Yonemoto (Editors)

Formation and Structural Change of Cake during Crossflow Microfiltration of Microbial Cell Suspension Containing Fine Particles

Takaaki TANAKA[a], Nobuyoshi YAMAGIWA[b], Tetsuya NAGANO[b], Masayuki TANIGUCHI[a], Kazuhiro NAKANISHI[b]

[a]Department of Material Science and Technology, Niigata University,
2-8050 Ikarashi, Niigata 950-2181, Japan
[b]Department of Bioscience and Biotechnology, Okayama University,
3-1-1 Tsushima-Naka, Okayama 700-8530, Japan

The permeation behavior in crossflow filtration of a model suspension containing yeast cells and polystyrene latex particles was studied. When the membrane pore size was larger than the latex particle size, a yeast cell layer was first formed on the membrane surface and then latex particles were deposited on it. When the pore size was smaller than the latex particle size, a yeast cell layer containing latex particles was formed on the membrane surface at the start of filtration. However, it was then replaced by the latex particle layer.

1. INTRODUCTION

Permeation flux decreases with the formation of a microbial cell layer (cake) during crossflow filtration (CFF) of microbial broths (1). Furthermore, the decrease of the permeation flux is accelerated when the broth contains fine particles derived from medium components and polymers secreted by the cells even in a small quantity. We have shown that the second layer of the fine particles or polymers formed on the cake surface causes the increase in the resistance of the cake more than 10 times (2, 3). The resistance of membrane is also increased by plugging of the membrane pores with fine particles (2). The insoluble salt formed during autoclave and the antifoams were also reported to decrease the permeation flux in crossflow filtration (4, 5).

In this paper, we cross-filtered a suspension, containing yeast cells and fine latex particles as a model broth with membranes having different pore sizes. We investigated the behavior for cake formation with respect to the permeation flux. We observed the structural change of the cake during crossflow filtration using a scanning electron microscope.

2. MATERIALS AND METHODS

2.1. Membranes

Four screen filters (pore size: 0.20, 0.45, 0.80, and 3.0 μm; Advantec, Co., Tokyo) and two depth filters (pore size: 0.22 and 3.0 μm; Fuji Photo Film, Co., Tokyo) were used. These membranes were all made of cellulose acetate.

2.2. Suspension

As a model of microbial broths, a suspension containing baker's yeast cells and polystyrene latex was used. The baker's yeast was cultivated for 24 h in a medium containing 10 kg m^{-3} yeast extract (Difco Laboratories, Detroit, MI), 20 kg m^{-3} polypepton (Nippon Pharmaceuticals, Tokyo), and 20 kg m^{-3} dextrose of pH 7.0. Yeast cells were washed by 0.9% NaCl solution and suspended in the same solution (6, 7). The cell sizes were (5.3±1.2) x (4.7±1.2) μm, which was measured by optical microscopy. The polystyrene latex was synthesized by a suspension polymerization method (8). The size was determined to be 0.50±0.05 μm by scanning electron microscopy. The latex suspension was used for the filtration experiments after dialysis against purified water. The concentrations of yeast and latex particles were 20 kg m^{-3} (in wet weight) and 0.8 kg m^{-3}, respectively.

2.3. Crossflow Filtration

A thin-channel type module was used. The length, width, and depth were 100, 24, and 2.4 mm, respectively. The filtration was performed at 20°C, with a transmembrane pressure of 49 kPa and circulation flow rate of 30 cm^3 s^{-1}. The linear flow rate, shear stress, and Reynolds number was evaluated to be 0.52 m s^{-1}, 1.3 Pa, and 2300, respectively (9). The permeate was returned to the reservoir tank to keep the concentrations (Figure 1).

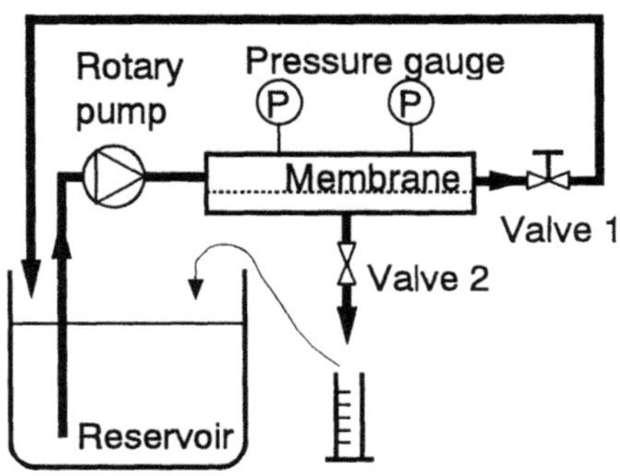

Figure 1. Schematic diagram for crossflow filtration experiments.

2.4. Scanning Electron Microscopy

The cross section of the filter cake formed on the membrane was observed by a scanning electron microscope (S-2150, Hitachi, Tokyo) after fixed with glutaraldehyde, lyophilized, and sputter-coated with Au-Pd.

3. RESULTS AND DISCUSSION

3.1. Effect of the addition of fine particles on the permeation flux in the crossflow filtration of a yeast cell suspension

A steady-state permeation flux of 1 x 10^{-4} m^3 m^{-2} s^{-1} was obtained at a filtration time of 600 s in the crossflow filtration of a 20 kg m^{-3} yeast cell suspension with a screen filter having 0.20 μm pores. When 0.8 kg m^{-3} of latex was added to the yeast suspension, the permeation flux reached nearly a steady state at 600 s. However, the flux was one sixth (1.5 x 10^{-5} m^3 m^{-2} s^{-1}) that without the latex particles.

3.2. Effect of the membrane pore size on the permeation flux

Figure 2 shows the behaviors of the permeation flux in the crossflow filtration with different screen filters, which indicates different behaviors of the permeation flux depending on the membrane pore sizes. The steady-state permeation fluxes with the membranes having pores larger (0.80 and 3.0 μm) than that of the latex particles (0.50 μm) were twice those with the membranes having pores smaller (0.20 and 0.45 μm) than that of the latex particles. A similar tendency was observed with depth filters (data not shown).

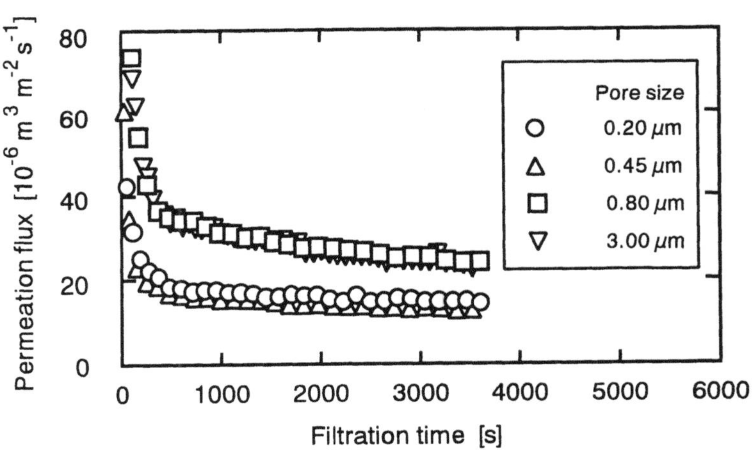

Figure 2. Effect of membrane pore size on the permeation flux.
The size of the latex particles was 0.50±0.05 μm

3.3. Formation of the filter cake in the crossflow filtration where the membrane pore size was larger than the latex

During the initial stage of the crossflow filtrtaion, the yeast cells (5 μm) deposited on the membrane and the latex particles (0.5 μm) passed through the both cell layer and membrane pores. As the permeation flux decreased the deposition of the yeast cells tended to be reduced. After the cease of deposition of the yeast cells, the latex particles started to deposit on the surface of the cell layer. The formation of a latex layer increased the permeation resistance of the cake to a great extent although it was very thin. Figure 3 shows the cross section of the cake formed on the membrane observed with a scanning electron microscope after 3600 s of the crossflow filtration with a depth filter having 3.0 μm. Even after the deposition of the latex, the both layers of the latex and yeast cells remained in the cake (Figure 4). These results were similar in tendency to the formation of the fine particle layer in the crossflow filtration of a yeast broth cultivated in molasses medium (2) and that of the polymer layer in the filtration of a *Corynebacterium glutamicum* broth (3).

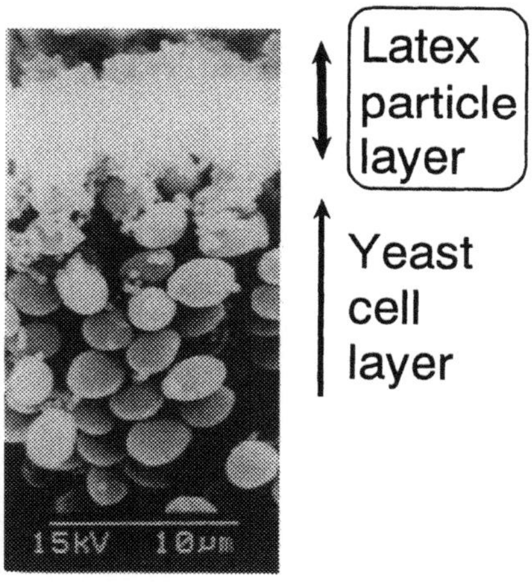

Figure 3. Cross section of the filter cake when the membrane pore was larger (3.0 μm) than that of the latex particles. The filtration time was 3600 s.

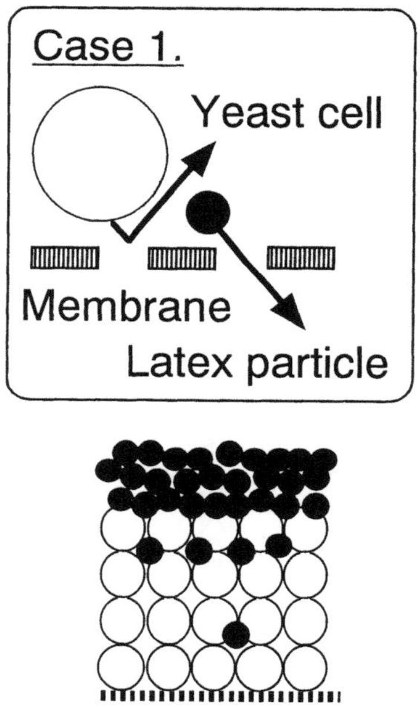

Figure 4. Formation of the cake when the membrane pore was larger than that of the latex particles.

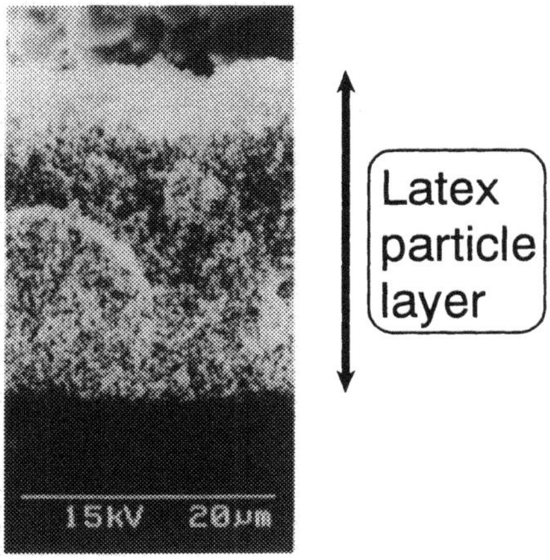

Figure 5. Cross section of the filter cake when the membrane pore was smaller (0.20 μm) than that of the latex particles. The filtration time was 3600 s.

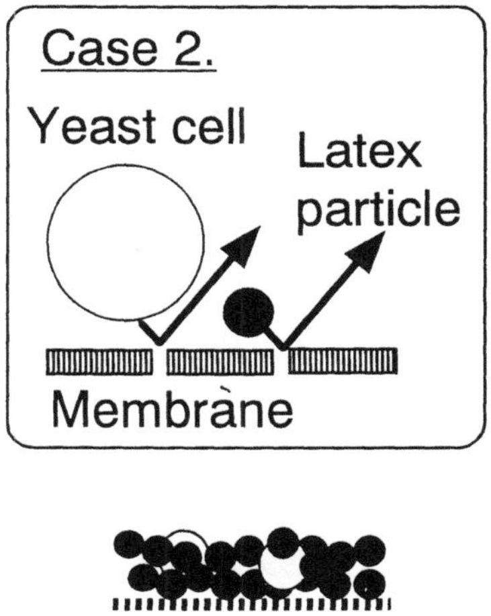

Figure 6. Formation of the cake when the membrane pore was smaller than that of the latex particles.

3.4. Formation of the cake in the crossflow filtration where the membrane pore size was smaller than the latex particle size

When a membrane with 0.20-μm pores was used, the both yeast cells and latex particles deposited on the membrane as a cake at the very initial stage. Then, the latex particles passed through the cell layer and reached to the membrane surface, which yielded the decrease in the permeation flux considerably. As a result, the yeast cells were swept by the shear of suspension flow parallel to the membrane and then the cell layer was completely replaced with a thin latex particle layer having a high permeation resistance (Figures 5, 6).

4. CONCLUSION

The steady-state permeation flux was considerably decreased by the addition of a small amount of fine latex particles in the crossflow filtration of yeast cell suspension. It was also shown that the cake formation depended on the membrane pore size relative to the latex particle size. These results would be helpful to understand the fouling of the membrane and to select the membrane in crossflow filtration of microbial broths.

REFERENCES

1. G. Belfort, R. H. Davis, and A. L. Zydney, J. Membrane Sci., 96 (1994) 1-58.
2. T. Tanaka, R. Kamimura, R. Fujiwara, and K. Nakanishi, Biotechnol. Bioeng., 43 (1994) 1094-1101.
3. T. Tanaka, K. Usui, and K. Nakanishi, Sep. Sci. Technol., 33 (1998) 707-722.
4. N. Nagata, K. J. Herouvis, D. M. Dziewulski, and G. Belfort, Biotechnol. Bioeng., 34 (1989) 447-466.
5. K. H. Kroner, W. Hummel, J. Volkel, and M.-R. Kula, pp. 223-232, in E. Dorioli and M. Nakagaki (eds.), Membranes and Membrane Processes, Plenum Press, New York, 1986.
6. T. Tanaka, R. Kamimura, K. Itoh, K. Nakanishi, and R. Matsuno, Biotechnol. Bioeng., 41 (1993) 617-624.
7. T. Tanaka, S. Tsuneyoshi, W. Kitazawa, and K. Nakanishi, Sep. Sci. Technol., 32 (1997) 1885-1898.
8. A. Kotera, K. Furusawa, and Y. Takeda, Kolloid-Zeitschrift und Zeitschrift Polymere, 230 (1970) 677-681.
9. H. Schlichting, pp. 612-615, Boundary-Layer Theory 7th ed., McGraw-Hill, New York, 1979.

Bioseparation Engineering
I. Endo, T. Nagamune, S. Katoh and T. Yonemoto (Editors)

Continuous separation of ternary mixture of amino acids using rotating annular chromatography with partial recycle of effluent

Takuya Fukumura[a], V. M. Bhandari, Akio Kitakawa[b] and Toshikuni Yonemoto[a]

[a]Dept. of Chem. Eng., Tohoku University, Aoba-yama 07, Sendai 980-8579, Japan

[b]Dept. of Materials Sci. & Eng., Miyagi National College of Technology, Natori 981-1239, Japan

Rotating annular chromatography with partial recycle of effluent (RAC-PRE) is applied to the separation of ternary mixture of amino acids. The mathematical model describing the separation process is constructed to find the reasonable operating condition. The effectiveness of RAC-PRE for the separation of a multicomponent mixture is experimentally and theoretically elucidated.

1. INTRODUCTION

Rotating Annular Chromatography (RAC) proposed by Martin[1] is a potential technique for continuous separation of a multicomponent mixture. In this technique, by continuously supplying the feed mixture from a fixed point of a rotating annular packed bed, each component elutes from its intrinsic position at the bottom of the bed determined by its retention strength to the chromatographic packing (i.e., the stronger the retention strength, the larger the elution angle from the feed point). This technique has been widely used in biochemical applications, such as protein separation by Hashimoto[2] and Bloomingburg[3] and amino acids separation by Goto[4] and Carta[5].

In conventional RAC operation, longer residence time of solute through the bed gives better resolution but simultaneously causes the industrially undesirable reduction of throughput. In order to overcome such a problem, we[6] have proposed RAC with Partial Recycle of Effluent (RAC-PRE) where effluents in which two components overlap are recycled to the inlet of the annular bed. The effectiveness of this technique was successfully proved by achieving the complete separation of a binary mixture of amino acids.

In industrial chromatographic separation processes, isolation of a target component from a multicomponent mixture is frequently desired. In most cases, elution profiles except that of the target component can be categorized into two individual groups. One group (group A) is made up of profiles of components having weaker retention strength to the packing than that of the target component and the other (group B) having stronger retention strength. Therefore, isolation of a target component from a multicomponent mixture can be reduced to the simultaneous separation of a ternary mixture (target component and groups A and B).

In this work, RAC-PRE is applied to the separation of a ternary mixture of amino acids. A mathematical model describing the ternary separation process is also constructed by modifying the model for the binary separation process reported previously[6,7]. The effectiveness of RAC-PRE for the separation of a ternary mixture is elucidated experimentally and theoretically.

2. PRINCIPLE OF RAC-PRE FOR THE SEPARATION OF TERNARY MIXTURE

Figure 1 shows the schematic diagram of the typical concentration profiles at both inlet (a) and outlet (b) of the bed in conventional RAC operation for the separation of a ternary mixture (C, D and E). As shown in (b), there are two incompletely separated elution regions, namely, Y (C+D) and Z (D+E).

Figure 2 shows the concentration profiles obtained in RAC-PRE together with recycling loops. Some recycling effluent nozzles are fixed adjacent to the raw feed nozzle above the bed. Among them, nozzles in the smaller angle than that of the raw feed nozzle are classified into groups I and II. II is positioned between I and the raw feed nozzle. Similarly in the larger angle, recycling effluent nozzles are classified into groups III and IV. III is positioned between the raw feed nozzle and IV. In region Y, effluents predominant in C over D are returned to recycling nozzles I and the others to III. On the other hand, in region Z, effluents predominant in D over E are returned to II and the others to IV. By such a partial recycling operation, each concentration profile at the inlet forms in the increasing order of the respective retention strength, so that each profile is in " semi-separated " shape. It enhances the separation performance compared with the conventional RAC operation.

3. MATHEMATICAL MODEL

The mathematical model is constructed by modifying the model reported in our previous work[6] on a binary separation of glutamic acid and valine using cation exchange resin packing.

Figure 3 shows the schematic diagram of the states in liquid and resin phases. Mass balance in liquid phase is expressed as equation(1) by considering axial and circumferential dispersion, convection, rotation and ion exchange between the liquid and resin phases.

$$E_{i,x}\frac{\partial^2 C_i}{\partial x^2}+E_{i,y}\frac{\partial^2 C_i}{\partial y^2}-u\frac{\partial C_i}{\partial x}-u_y\frac{\partial C_i}{\partial y}-\frac{3u_y(1-\varepsilon)}{R_p^{\,3}\varepsilon}\frac{\partial}{\partial y}\int_0^{R_p} q_i r^2 dr=0 \tag{1}$$

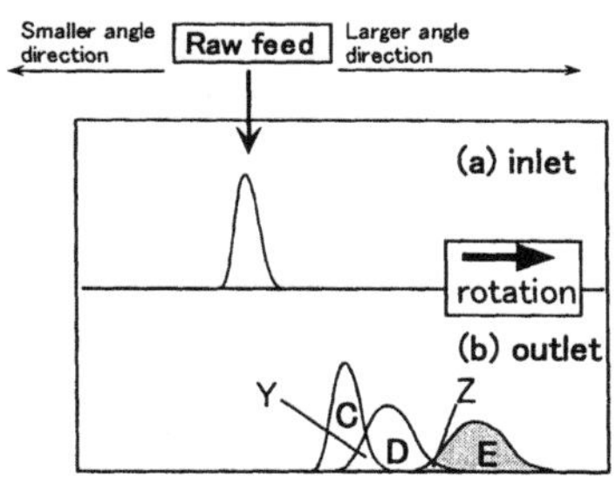

Fig.1 Concentration profiles in conventional RAC operation

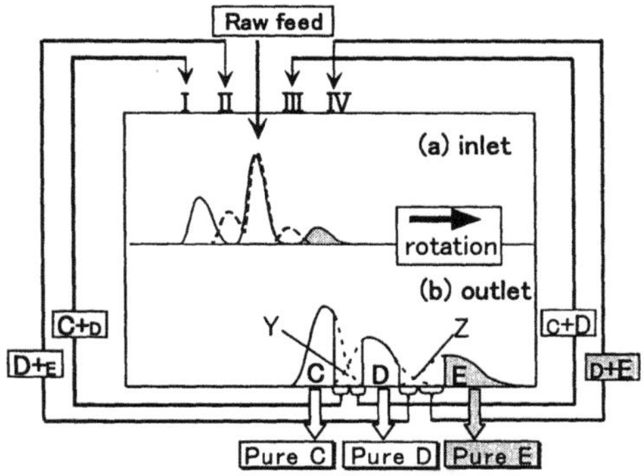

Fig.2 Concentration profiles in RAC-PRE operation

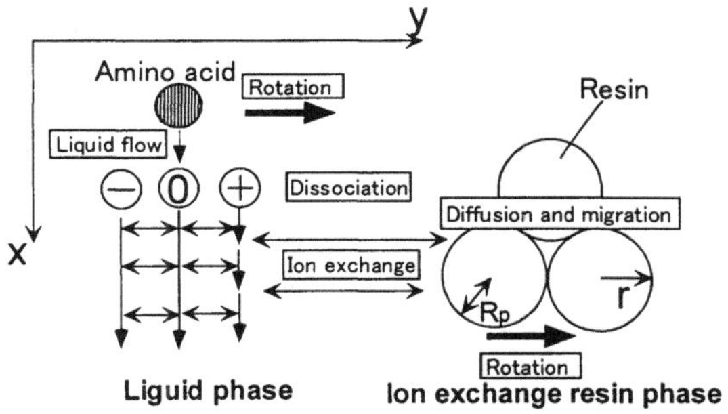

Fig.3 Schematic diagram of the states in liquid and resin phases

In resin phase, by considering the intraparticle diffusion and migration and the flux arising from rotation of the bed, the mass balance is expressed as,

$$u_y \frac{\partial q_j}{\partial y} = \frac{1}{r^2} \frac{\partial}{\partial r}\left(r^2 \sum_{\lambda} \overline{D_{j,\lambda}} \frac{\partial q_\lambda}{\partial r} \right) \tag{2}$$

Here,

$$\overline{D_{j,\lambda}} = \begin{cases} -\dfrac{\overline{D_j}(\overline{D_\lambda} - \overline{D_{H^+}})q_j}{\sum_{k=all\ cations} \overline{D_k} q_k} & (j \neq \lambda) \\ \overline{D_j} - \dfrac{\overline{D_j}(\overline{D_j} - \overline{D_{H^+}})q_j}{\sum_{k=all\ cations} \overline{D_k} q_k} & (j = \lambda) \end{cases} \tag{3}$$

and

i = Glu, Val, Leu, Acetic acid, Sodium ion, Hydrogen ion

j, λ = Glu, Val, Leu, Sodium ion

These equations combined with the auxiliary equations, such as the dissociation equilibrium relationships of the dissociative species, are numerically solved using the finite difference method under the conditions suitable for the partial recycling operation.

4. EXPERIMENTAL

Figure 4 shows the schematic diagram of RAC-PRE apparatus. The annular bed has the inner diameter of 150mm, the outer diameter of 160mm, the thickness of 10mm and the height of 290mm. It is rotated at a constant rate using a stepping motor controlled by a personal computer. The annular space is packed with the cation exchange resin UBK530 (Mitsubishi Chemical Co., 0.22mm in mean diameter, Na^+ type). 36 nozzles are fixed above the bed. One nozzle is used for the raw feed mixture and some

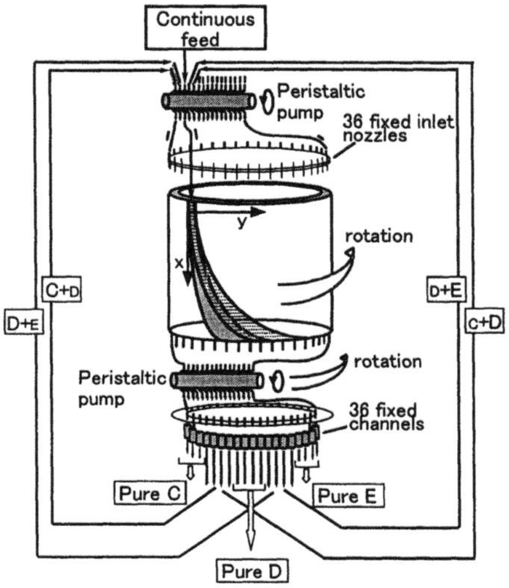

Fig.4 Schematic diagram of RAC-PRE apparatus

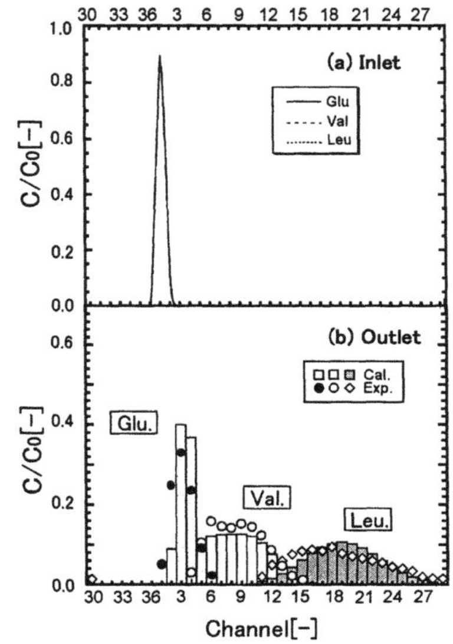

Fig.5 Concentration profiles without partial recycling

nozzles adjacent to that are used as recycling effluent nozzles. The others are used for the eluent solution. A sampler of 36 channels for the recovered effluents is fixed below the bed. The raw feed nozzle is labeled nozzle 1. The fixed channel just under the nozzle 1 is labeled channel 1. The numbers of the nozzles and channels are labeled in direction of rotation.

The concentration of each amino acid in the feed is prepared to be $1 mol/m^3$, respectively. The isoelectric points of the respective amino acids are 3.22 (Glu), 5.96 (Val) and 5.98 (Leu). Sodium acetate buffer solution is used as an eluent. Sodium concentration and pH of the eluent are adjusted to $8 mol/m^3$ and 4.9, respectively. The order of the retention strength of solutes is Glu < Val < Leu in this pH. Here, valine corresponds to the target component. The recycling channels are determined on the basis of the results of the preliminary experiment and the numerical simulation without partial recycling. After attaining steady state, the recovered effluents are quantitatively analyzed by HPLC.

5. RESULTS AND DISCUSSION

5.1. Comparison between conventional RAC and RAC-PRE

Figure 5 shows the preliminary experimental and calculated results of conventional RAC operation without partial recycling at superficial liquid velocity of 1.34×10^{-3}m/s and rotation rate of 0.065deg/s. The ordinate represents the concentration of amino acid nondimensionalized by its feed concentration and the abscissa represents the channel number. Figure (a) shows the calculated result at the inlet of the bed and (b) the results at the bottom. Amino acids were not completely separated in channels 4-6 and channels 11-17, respectively. Incompletely separated effluents can be categorized into 4 groups of

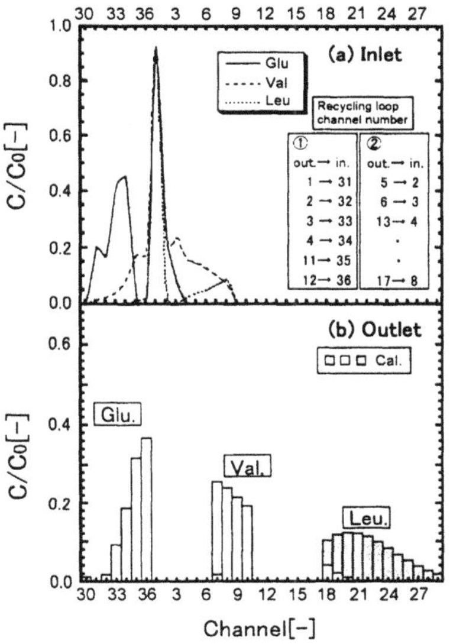

Fig.6 Concentration profiles with partial recycling

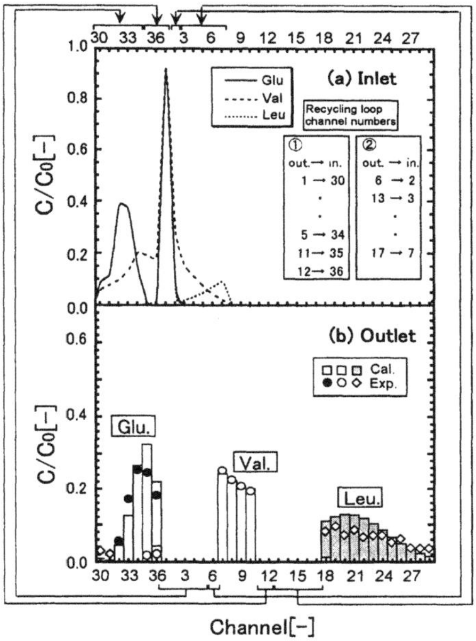

Fig.7 Concentration profiles with modified partial recycling

channels, 1-5, 6, 11-12 and 13-17 based on their respective predominant components. Following the recycling principle mentioned above, these groups were recycled to nozzles, 31-34, 2-3, 35-36 and 4-8 as Ⅰ, Ⅱ, Ⅲ and Ⅳ, respectively.

Figure 6 shows the calculated results with partial recycling. As shown in (b), partial recycling operation obviously improved the separation performance compared with Fig.5(b) because the numbers of incompletely separated channels were reduced and the concentration of valine doubled. However, at the inlet, the concentration profile of glutamic acid in larger angle than that of the raw feed nozzle was relatively broad because two channels 5, 6, in which glutamic acid eluted, were recycled to the nozzles 2 and 3. It caused the contamination in the elution channel of valine (channel 7). The recovery of isolated valine should be improved.

5.2 Modification of recycling loops

Recycling loops were rearranged for the purpose of higher recovery of isolated valine. Channel 6 was returned to the recycling nozzle (nozzle 2) in order to prevent the glutamic acid from eluting in channels in larger angle than that of channel 7. Together with such a change of a recycling loop, recycling nozzles corresponding to the other recycling channels were shifted by one position in reverse direction of rotation.

Figure 7 shows the results with partial recycling using the modified recycling loops. As shown in (a), glutamic acid in larger angle than that of the raw feed nozzle disappeared. As shown in (b), valine and leucine were completely separated. Furthermore, the calculated results were in reasonable agreement with the experimental results. It was found that the simulation technique could favorably help us to predict the actual separation process.

6. CONCLUSION

RAC-PRE for the separation of a ternary system was constructed on the basis of the previous work on a binary system. By applying this technique to the separation of the ternary mixture of amino acids, the separation performance was significantly improved compared with conventional RAC operation. Furthermore, the calculated results based on the mathematical model constructed by modifying the model reported in our previous work could reasonably predict the actual separation process. The effectiveness of RAC-PRE for the separation of a multicomponent mixture was elucidated experimentally and theoretically.

NOMENCLATURE

C	:	Concentration in liquid phase
$\bar{D}$	:	Effective diffusivity
E	:	Dispersion coefficient
q	:	Concentration in resin phase
R_p	:	Radius of resin
u	:	Superficial liquid velocity
u_y	:	Rotation rate
x	:	Axial coordinate in direction of liquid flow
y	:	Circumferential coordinate
r	:	Radial coordinate in resin phase

Greekss

ε	:	Void fraction

REFERENCES

1. A. J. P. Martin, *Discuss. Faraday Soc.*, **7** (1949) 332.
2. K. Hashimoto, M. Morishita, S. Adachi, K. Shindo, Y. Shirai and M. Tanigaki, *Preparative Chromatography*, **1** (1989) 163.
3. G. F. Bloomingburg, J. S. Bauer, G. Carta and C. H. Byers, *Ind. Eng. Chem. Res.*, **30** (1991) 1061.
4. Y. Takahashi and S. Goto, *Sep. Sci. Technol.*, **25** (1990) 1131.
5. C. H. Byers and G. Carta, *AIChE J.*, **36** (1990) 1220.
6. A. Kitakawa, Y. Yamanishi and T. Yonemoto, *Ind. Eng. Chem. Res.*, **36** (1997) 3809.
7. A. Kitakawa, T. Yonemoto and T. Tadaki, *Trans. Inst. Chem. Eng., Part C*, **72** (1994) 201.

Bioseparation Engineering
I. Endo, T. Nagamune, S. Katoh and T. Yonemoto (Editors)

Mass Transfer Characteristics of a Perfusion-type gel Analyzed by Shallow Bed Method

Masaaki Terashima, Shinji Nishimura, and Hiroyuki Yoshida

Department of Chemical Engineering, College of Engineering,
Osaka Prefecture University,
1-1 Gakuen-Cho, Sakai, Osaka 599-8531, Japan

Abstract

Up-take curve of POROS HQ50 for protein depended on the flow rate of BSA solution, while that of Sepharose FF was independent of the flow rate. This result indicates that adsorption rate of the perfusion-type gel (POROS HQ50) is affected by the intra-particle convective flow. Intra-particle effective diffusivity of Sepharose FF increased with the increase in BSA concentration. Parallel diffusion model revealed that the pore diffusion was dominant in Sepharose FF. On the other hand, the intra-particle effective diffusivity in POROS HQ50 did not depend on BSA concentration. Surface diffusion was dominant in POROS HQ50, and the surface diffusivity depended on the flow rate due to the intra-particle convection.

Key words: perfusion, protein transport, surface diffusion, intra-particle convection

1. Introduction

While the significance of convective transport was first described in catalytic gas reaction, Afeyan and coworkers introduced perfusion chromatography, taking advantage of intra-particle convection, to HPLC system in 1989 (1). Although many reports demonstrated the advantages of perfusion-type gel, and explained the intra-particle mass transfer mechanism (2, 3), it is still far from the substantial understanding of the perfusion phenomena. For this reason, we have characterized the intra-particle mass transfer of a perfusion-type gel using shallow bed method, and obtained effective diffusivities were analyzed by the parallel diffusion model (4). POROS HQ50 (Perseptive Biosystems) and DEAE Sepharose FF (Amersham Pharmacia Biotech) were used as a perfusion gel and a reference gel, respectively. The shallow bed method and its theoretical back ground were described elsewhere (4).

2. Theory

2.1 Adsorption equilibrium

Equilibrium isotherms for BSA adsorption on perfusion type gel were correlated by the Langumuir equation (Eq. 1).

$$q=\frac{QK_EC}{1+K_EC} \tag{1}$$

where, q [kg/m^3-wet resin], C [kg/m^3], and K_E [m^3/kg] are solid phase equilibrium BSA concentration, liquid phase BSA concentration, and Langmuir constant, respectively.

2.2 Estimation of Intra-particle Effective Diffusivity

In order to determine intra-particle effective diffusivity from experimental data, we have assumed homogeneity of the gels. Assuming Fickian diffusion, fractional attainment of equilibrium (F) to correlate kinetic data was given by Eq. 2 (5):

$$F=1-\frac{6}{\pi^2}\sum_{n=1}^{\infty}\frac{1}{n^2}\exp(-D_{eff}n^2\pi^2t/r_0^2) \tag{2}$$

where, F[-], D_{eff} [m2/s], t [s], r_0 [m] are fractional attainment of equilibrium, intra-particle effective diffusivity, time, and gel radius, respectively.

2.3 Parallel Transport model by Surface and Pore Diffusion

Mass balance equation of parallel transport model in the particle is given by the following equation (Eq. 3).

$$\varepsilon\frac{\partial C}{\partial t}+\frac{\partial q}{\partial t}=D_P\varepsilon\frac{1}{r^2}\frac{\partial}{\partial r}(r^2\frac{\partial C}{\partial r})+D_s\frac{1}{r^2}\frac{\partial}{\partial r}(r^2\frac{\partial q}{\partial r}) \tag{3}$$

where, D_P [m^2/s], D_s [m^2/s], ε [-] are pore diffusivity, surface diffusivity, and porosity of gels, respectively.

According to Yoshida *et al.* (4), the relationship among the effective intra-particle diffusivity D_{eff}, the surface diffusivity D_s and approximate pore diffusivity D_{pa} is given by the following equation (Eq. 4).

$$D_{eff}(1+\frac{1}{\alpha}) = D_S + D_{pa}\frac{1}{\alpha} \tag{4}$$

where, D_{pa} [m^2/s] is approximate pore diffusivity , and $\alpha=\varepsilon C_0/q_0$ [-].

This equation shows that the surface diffsivity and approximate pore diffsivity can be obtained by plotting $D_{eff}(1+1/\alpha)$ against $1/\alpha$. The slope of the equation 4 gives approximate pore diffusivity due to an assumption made to derive this equation (4).

3. Materials and Methods

3.1 Perfusion-type gel

POROS HQ50 (Perseptive Biosystems) and DEAE Sepharose FF (Amersham Pharmacia Biotech) were used as a perfusion gel and a reference gel, respectively. Characteristics of these gel are summarized in Table 1. Bovine serum albumin (BSA) was used as an adsorbent.

3.2 Shallow-bed method

In order to determined the intra-particle effective diffusivity by Eq. 2, we measured kinetic data with the shallow bed method. As shown in Fig. 1, small amount of the gel was packed in the column (inner diameter, 1 cm or 0.6 cm). A phosphate buffer (20 mM, pH 6.9) was fed into the column to equilibrate the gel. The phosphate buffer containing BSA was instantaneously fed to the column for predetermined time at high flow rate so that the BSA concentration at the gel surface is constant. After washing the gel with the equilibrium buffer for 20 second, the adsorbed BSA was eluted with the phosphate buffer containing 1 M NaCl. The mount of adsorbed BSA was determined by measuring the absorbance at 280 nm with a specrtrophotometer (Shimadzu SPD-6A). Fractional attainment of the equilibrium was obtained by dividing the amount of the adsorbed BSA by the equilibrium value calculated from Langumuir isotherms. Protein flow rate was changed from Re=0.0054 to Re=2.4, and BSA concentration was changed from 0.5 kg/m^3 to 5.0 kg/m^3. All experiments were carried out at 25 °C.

4. Results and Discussion

Parameters of Langumuir isotherm for the both gels are summarized in Table 2. The Langumuir isotherm well explained the experimental data for both gels.

Fig. 2 shows the effect of the flow rate of the protein solution on BSA up-take curve for Sepharose FF at BSA concentration 0.5 kg/m^3. In the case of Sepharose FF, the flow rate of the external solution did not affect the up-take curve, and the up-take curve was well characterized by Eq. 4. This results show that intra-particle transport of BSA is governed by diffusion, and the intra-particle convective flow did not occur in Sepharose FF gel. On the other hand, the flow rate of the protein solution strongly affected the up-take curve in the case of POROS HQ50 as shown in Fig. 3. The intra-particle diffusivity, obtained from the up-take curve, increased with the increase in the flow rate of the external solution. This result indicates that the adsorption rates of the perfusion-type gel are affected by the intra-particle convective flow.

Fig. 4 shows the relation ship between $D_{eff}(1+1/\alpha)$ and $1/\alpha$ for Separose FF at Re=0.61. According to the parallel transport model and Eq. 4, rate limiting step of BSA transport in the Sepharose FF is pore diffusion. Fig. 5 shows similar plots for POROS HQ50. This results suggests that the rate limiting step of BSA transport in the gel is surface diffusion. It should be noted that the surface diffusivity of POROS 50 strongly depend on the flow rate as shown in Fig. 6.

Although adsorption rates of Sepharose FF were independent of the flow rates of bovine serum albumin (BSA) solution, those of POROS HQ50 increased with the increase of the flow rate. Theses results clearly show the difference of the perfusion-type gel and the non-perfusion-type gel.

Intra-particle effective diffusivity in the Sepharose FF were increased with the increase in BSA concentration (Fig. 4). Parallel diffusion model revealed that the intra-particle pore diffusion was dominant in Sepharose FF. On the other hand, the intra-particle effective diffusivity in the POROS HQ50 did not depend on BSA concentration (Fig.5). The parallel diffusion model shows that the surface diffusion was dominant in POROS HQ50, and the surface diffusivity strongly depends on the flow rate (Fig. 6). Thus, the effect of intra-particle convection of POROS HQ50 was described as the surface diffusion in the parallel diffusion model. The reason why the surface diffsivity depends on the flow rate can be explained as follows: Since diffusion rate of BSA into micro-particles which form a perfusion gel is rapid, and the mass transfer of BSA by intra-particle convection was insufficient under low flow rates, the average concentration of BSA at the surface of the micro-particles is lower than that at the surface of the perfusion gel. As the flow rate increases, the surface diffusivity increases because the average concentration of BSA at the surface of the micro-particles approaches to that at the surface of the perfusion gel. Finally, the surface diffusivity might become constant when the average concentration of BSA at the surface of the micro-particles is equal that at the surface of the perfusion gel.

5. Conclusion

Intra-particle convective flow in the perfusion-type gel was characterized by enhancement of protein up-take time in shallow bed method. This effect, analyzed by parallel transport model, could be expressed as surface diffusion in the gel. The surface diffusivity of the perfusion gel increased with increase in the protein flow rate.

References

(1) N.B. Afeyan, Proceedings of the ISPPP-89 meeting held in Philadelphia, PA, USA. (1989)
(2) S. Katoh, M. Terashima, E. Sada, H. Utsumi, Y. Kamiya, K. Yamada, T. Majima, J. Ferment. Bioeng., 78, 246-249 (1994)
(3) M. McCoy, K. Kalghatgi, F. E. Regnier, N. Afeyan, J. Chromatogr. *A*, 743, 221-229 (1996)
(4) H. Yoshida, M. Yoshikawa, T. Kataoka, AIChE J., 40, 2034-2044 (1994)
(5) Carslaw, H. S. and J. C. Jaeger, Conduction of Heat in Solids, Oxford, 2nd ed., pp. 91 (1975)

Table 1 Characteristics of gels

	Diameter (μm)	Porosity (-)
Separose FF	68.4	0.84
POROS HQ50	46.0	0.6

Table 2 Langmuir parameters of gels

	K_E (m^3/kg)	Q (kg/m^3-wet resin)
Separose FF	2.98	157
POROS HQ50	1.96	177

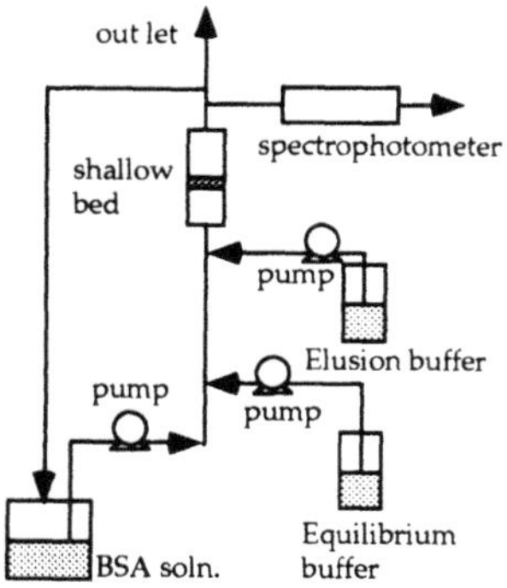

Fig.1 Experimental apparatus

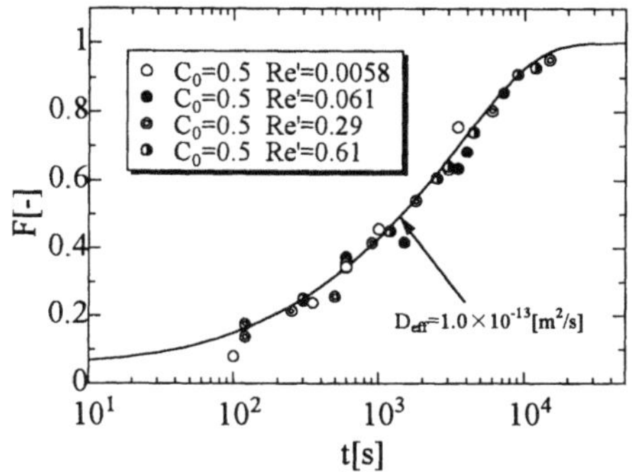

Fig.2 Up-take curve of Sepharose FF
(Re:○ 0.0058, ● 0.061, ◎ 0.29, ◐ 0.61)

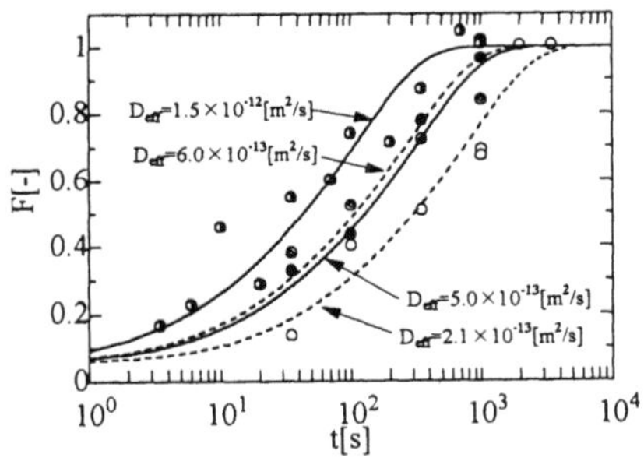

Fig. 3 Up-take curve (POROS)
(Re:○ 0.0054, ● 0.060, ◎ 0.61, ◐ 2.4)

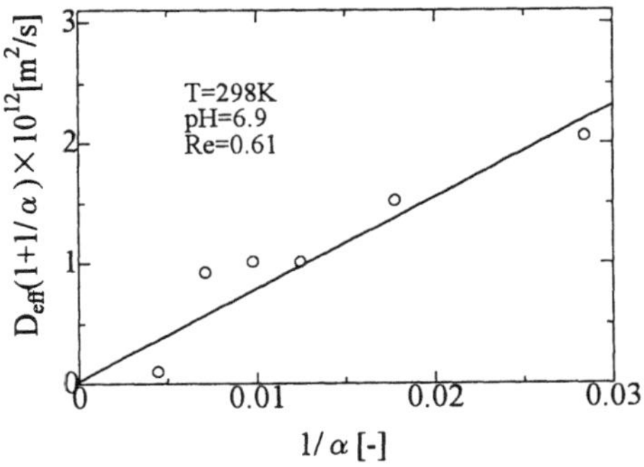

Fig. 5 $D_{eff}(1+1/\alpha)$ vs. $1/\alpha$ (Sepharose FF)

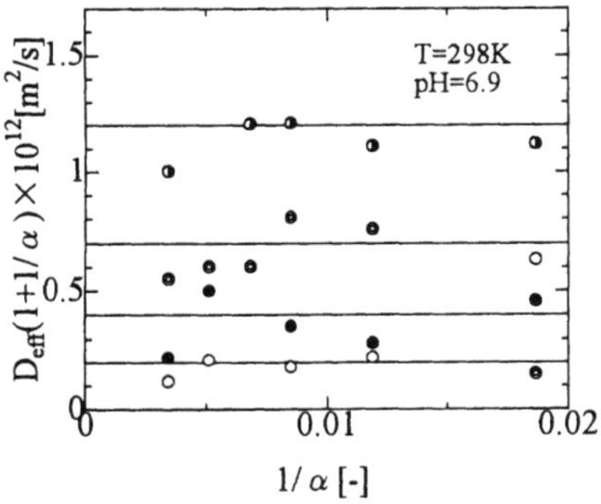

Fig. 5 $D_{eff}(1+1/\alpha)$ vs. $1/\alpha$ (POROS)
(Re:○ 0.0054, ● 0.060, ◎ 0.61, ◐ 2.4)

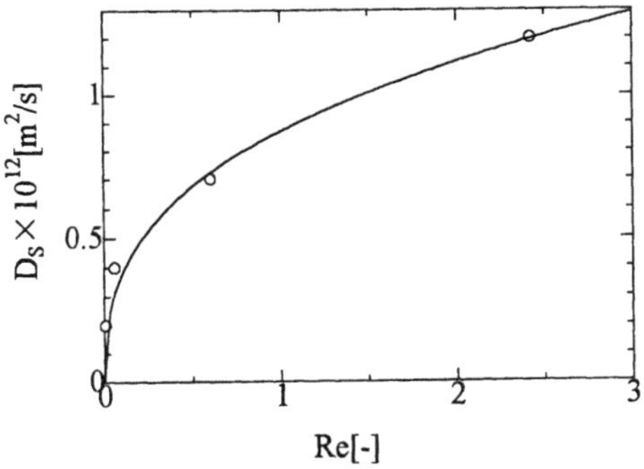

Fig. 6 Effect of flow rate on D_s (POROS)

Bioseparation Engineering
I. Endo, T. Nagamune, S. Katoh and T. Yonemoto (Editors)

Fouling of Cheese Whey during Reverse Osmosis and Precipitation of Calcium Phosphate

Hideki TSUGE*, Yuko TANAKA and Noriko HISAMATSU

Department of Applied Chemistry, Keio University,
3-14-1, Hiyoshi, Kohoku-ku, Yokohama 223-8522, Japan.

The effect of pH on flux and solute permeability of cheese whey and model solutions during reverse osmosis (RO) filtration was examined. The calcium phosphate was observed on the RO membrane during RO filtration process, whose morphology and shape were compared with the morphology profile and SEM photographs of calcium phosphate obtained by precipitation experiments at the conditions of reaction temperature 15- 50°C and the initial pH 5-9.

1. INTRODUCTION

For the effective usage of cheese whey, reverse osmosis (RO) membrane separation process has been applied in dairy industry with developing membrane technology. During membrane separation of whey, however, the fouling of the membrane causes the reduction of efficiency due to decrease in permeate flux and shortening membrane life.

Calcium phosphate is one of the most important inorganic constituents in whey, whose average contents of calcium and inorganic phosphate are 40 and 116mg/100g-serum, respectively[1]. As the whey is saturated with respect to calcium phosphate, a large part of calcium phosphate is undissolved. This means whey is thermodynamically unstable system and the precipitation of calcium phosphate may take place sooner or later according to operating conditions.

The aims of this study are as follows;

1. to make clear the effect of pH on flux and permeability of protein and calcium in whey during RO filtration. Whey, mixed solution of whey and ethylenediaminetetraacetic acid (EDTA), and whey protein isolate (WPI) were used as solutions to examine the combined effect of whey protein and calcium phosphate.
2. to clarify the effects of pH and temperature on the precipitation process of calcium phosphate in the batch system at the conditions of reaction temperature 15- 50°C and the initial pH 5-9.
3. To compare the morphology and shape of calcium phosphate obtained on RO membrane with the morphology profile and SEM photographs by precipitation experiments.

* e-mail: tsuge@applc.keio.ac.jp TEL:045-563-1141 Ex 3403 FAX:045-563-0446

2. EXPERIMENTAL

2.1. Reverse Osmosis Experiments

The experimental apparatus in recirculation mode was used for RO filtration as shown in Figure 1. The RO membrane, Nanomax 50 (Millipore Co.), was used, whose molecular weight cut off is 400 and the nominal rejection of sodium chloride is 50%. The effective membrane area was 4.53 $\times 10^{-3} m^2$. The pH of cheese whey and other solutions was adjusted from 4.0 to 8.0 by using NaOH or HCl with a digital pH meter[HM-40V, TOA Electronics Co.]. The feed was then maintained at 25°C about 1 hour. The prepared solution was stored into the 200mL tank and the RO unit was operated with 2.94MPa by N_2 cylinder. The impeller was operated at 800 rpm by the magnetic stirrer. The permeate was weighed by the balance.

Pure water flux was measured before all experiments until a constant flux was obtained under the above conditions to keep the initial flux of the RO membrane constant. The concentrations of calcium and phosphate ions were measured by the ion chromatography [IC-7000P, Yokokawa Co.].

2.2. Precipitation Experiments

The crystallizer was a 1-liter stirred tank reactor made of acrylic resin of 10cm diameter and was placed in a thermostat bath maintained at 15, 25, 38 or 50°C. The 4-blade turbine type impeller was operated at 357 rpm to ensure complete mixing.

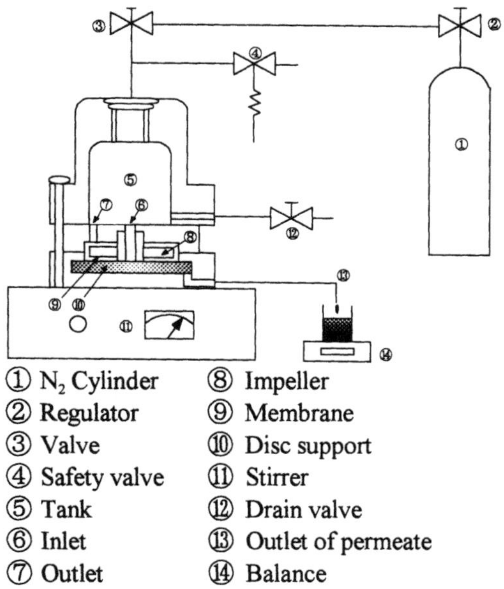

Figure 1. Experimental apparatus of RO filtration process

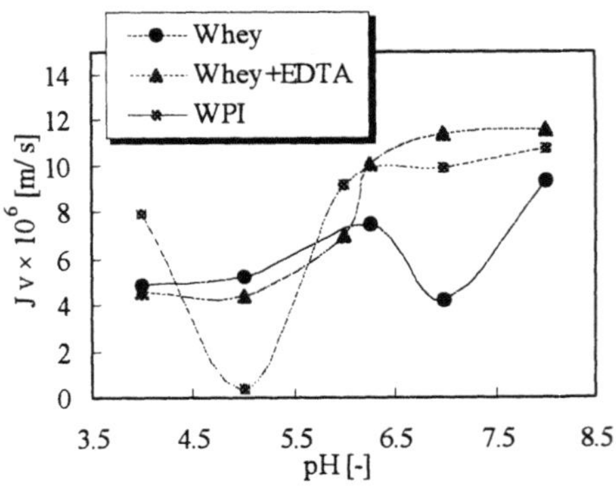

Figure 2. Relation between steady state flux Jv and pH

The precipitation of calcium phosphate was carried out by mixing of potassium phosphate and calcium nitrate aqueous solutions. The potassium phosphate solution was prepared by mixing of K_2HPO_4 and KH_2PO_4 and the pH of the solution was controlled by varying R, which is defined as the mole fraction of K_2HPO_4 to total potassium phosphate, K_2HPO_4+ KH_2PO_4. The concentration of calcium nitrate was changed from 4.0 to 16.0 mmol/dm^3 and R from 0.05 to 1.0. The calcium/phosphate molar ratio, Ca/P, was 1.0. In order to have the same ionic strength, I = 100 mmol/dm^3, potassium nitrate was added to phosphate aqueous solution before mixing calcium nitrate. At the end of the precipitation, the suspended solution sampled from the crystallizer was filtered by the microfilter with pore size 0.3 μm. After the natural drying, the particles were analyzed by X-ray diffraction and observed by SEM.

3. RESULTS AND DISCUSSION

3.1. Effect of pH on the flux

The compositions of cheese whey and WPI are shown in Tables 1 and 2. Figure 2 shows the relation between steady state permeate flux J_v of whey, mixed solution of whey and EDTA and WPI, and pH. The permeate flux of whey shows a maximum near pH 6. For pH<6, the permeate flux decreases with decrease in pH by the adsorption of whey proteins, while pH>6 the permeate flux decreases with increase in pH due to the precipitation of calcium phosphate.

The WPI solution shows a minimum nearly at pH 4.6, which corresponds to isoelectric point of whey protein. When EDTA(0.2mol/L) was added to the whey, the flux shows a similar tendency to the WPI for pH>6, which suggests that the precipitation of calcium phosphate is controlled by adding EDTA to whey.

Table1 Composition of cheese whey

Constituents	[wt%]
Water	93.91
Ash	0.532
Whey lipids	0.75
Whey proteins	0.809
Lactose	4.856

Table2 Composition of WPI*,**

Constituents	[wt%]
Water	3.23
Whey lipids	0.40
Whey proteins	91.90
α-lactalbumin	27.64
β-lactoglobulin	59.41
Bovine serum albumin	2.88
others	0.09
Non-protein nitrogen	1.85
Lactose	2.30
Ash	1.84

* : All of these data are analytical values, so the whole summation of these data is not 100% exactly, which is due to analytical error.
** : These data were presented by Snow Brand Co. Ltd.

3.2. Permeability of Ca^{2+} and PO_4^{3-}

Figures 3 and 4 show the relations between permeability of Ca^{2+} and PO_4^{3-} after RO filtration process of whey and mixed solution of whey and EDTA, and pH. From Figure 3, permeabilities of both ions decrease for pH≧6 due to the precipitation of calcium phosphate at the membrane surface. On the other hand, Figure 4 shows the permeabilities of both ions increase with increase in pH because of the chelate formation between EDTA and Ca^{2+}, and the retardation of calcium phosphate precipitation.

Figures 5 and 6 show the relations between permeability of Ca^{2+} and PO_4^{3-} after RO filtration process of mixed solution of WPI and calcium phosphate and mixed solution of WPI, calcium phosphate and EDTA, and pH. By adding EDTA, the permeabilities of both ions increase due to the chelate formation between EDTA and Ca^{2+} for pH>6.

3.3. Precipitation of calcium phosphate

Figure 7 shows the typical SEM photograph of calcium phosphate obtained on the RO membrane after the crossflow filtration of whey by RO apparatus shown in Figure 1 at the conditions t = 25 °C and pH =6 -7. By X-ray diffraction of calcium phosphate and membrane itself shown in Figure 8, both of octacalcium phosphate (OCP,$Ca_8H_2(PO_4)_6 \cdot 5H_2O$) and dicalcium phosphate dihydrate(DCPD, $CaHPO_4 \cdot 2H_2O$) were precipitated on the RO membrane surface.

3.4. Precipitation Diagram

From the precipitation experiments, the morphology profile of calcium phosphate was obtained as a function of temperature and initial pH at 1 hour after pH becomes steady state as shown in

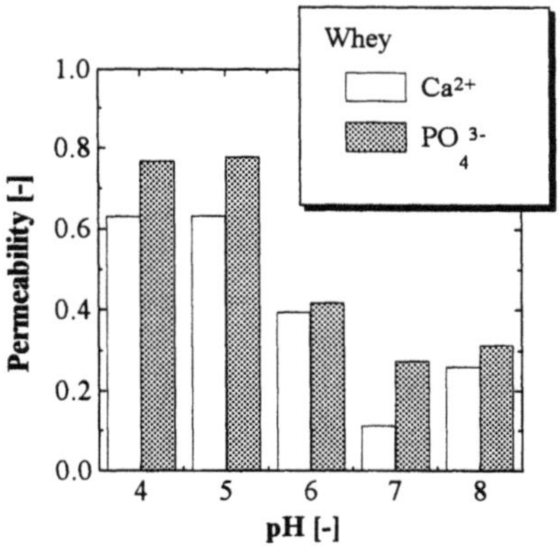

Figure 3. Relation between permeability and pH for whey

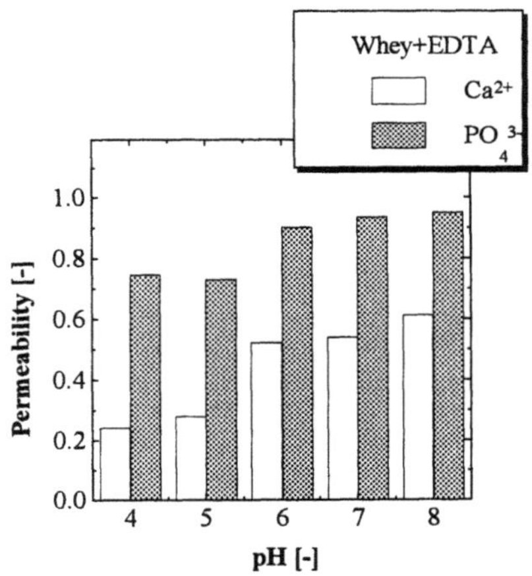

Figure 4. Relation between permeability and pH for whey+EDTA

Figure 9. The profile shows that pure DCPD was crystallized at low pH and low temperature, and conversely pure hydroxyapatite (HAP, $Ca_5OH(PO_4)_3$) was crystallized at high pH and high temperature. At the intermediate condition, OCP mixed with DCPD or HAP was obtained. As OCP is metastable phase, it is difficult to obtain pure OCP in batch system, where pH changes remarkably. At about pH<5.5, calcium phosphate was not formed.

Figure 10 shows SEM photographs of the observed calcium phosphate. (A) is DCPD, which is tabular, and (B) is OCP obtained with DCPD, which is spherical agglomerated particle composed of tabular crystals and similar to OCP itself. The crystal shape of calcium phosphate filtrated on the RO membrane is very similar to that shown in (B).

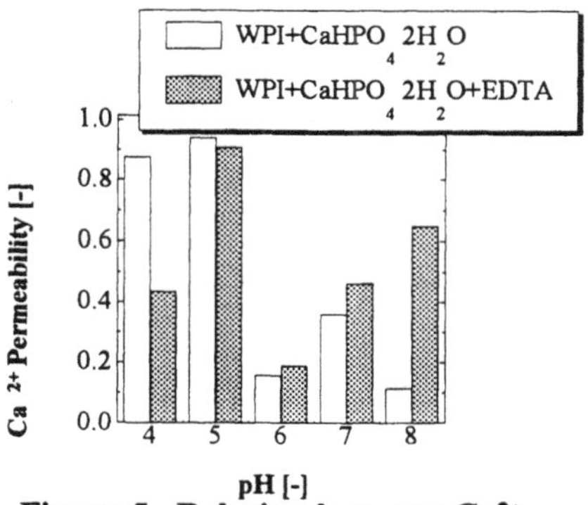

Figure 5. Relation between Ca^{2+} permeability and pH

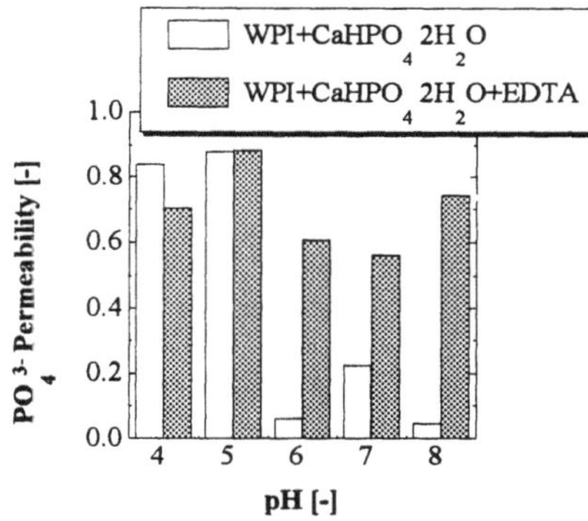

Figure 6. Relation between PO_4^{3-} permeability and pH

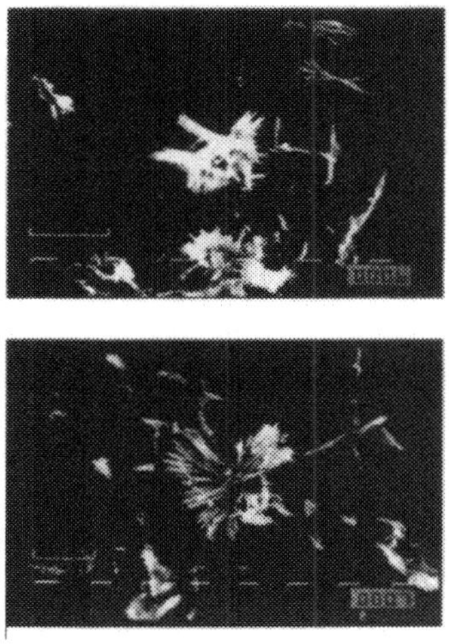

Figure 7. SEM photograph of calcium phosphate on RO membrane

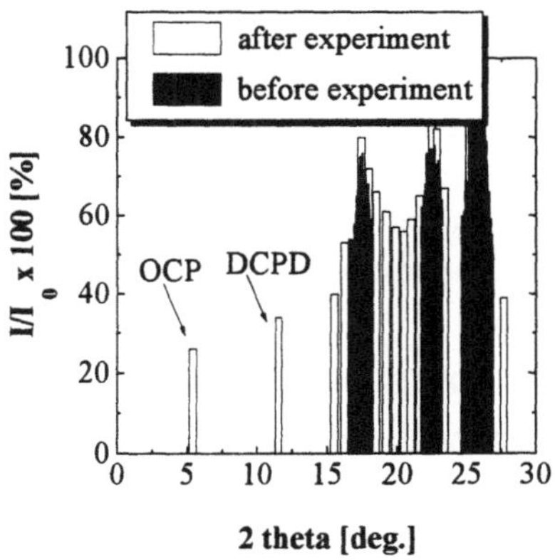

Figure 8. X-ray diffraction chart of calcium phosphate and membrane

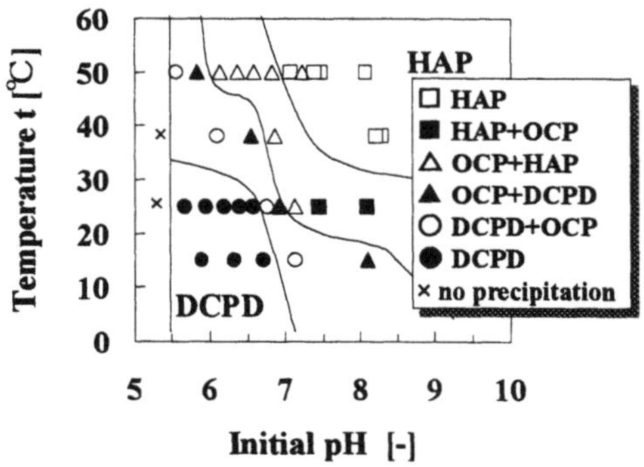

Figure 9. Precipitation profile of calcium phosphate

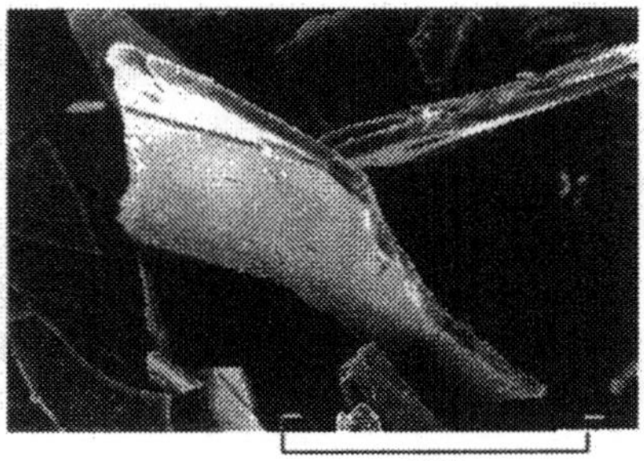

(A) DCPD

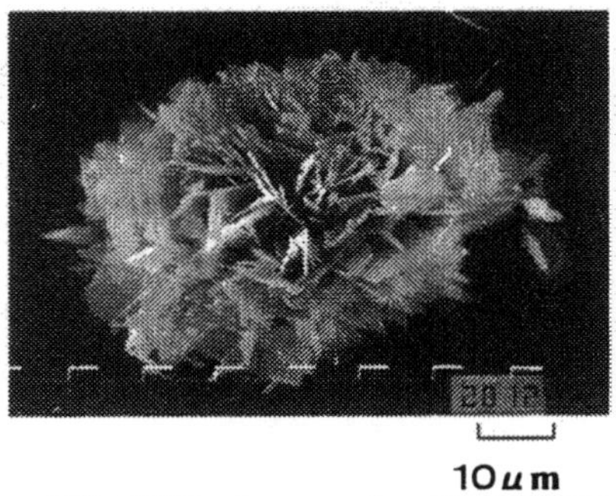

(B) OCP observed with DCPD

Figure 10. SEM Photographs of calcium phosphate

The addition of WPI affects the precipitation of calcium phosphate, so that its crystal shape changes by the addition of WPI. Further investigation should be continued.

4. CONCLUSIONS

1. The crossflow filtration of whey and model solutions with RO membrane was carried out. The pH condition of whey is one of the most remarkable causes for decreasing the permeate flux. Around pH=6, the permeate flux decreases due to the precipitation of calcium phosphate. The precipitation of calcium phosphate is controlled by adding EDTA to whey.
2. The calcium phosphate, that is, octacalcium phosphate and dicalcium phosphate dihydrate, was observed on the RO membrane surface. It was compared with the morphology profile of calcium phosphate obtained experimentally as a function of temperature and initial pH.

ACKNOWLEDGEMENT

Financial and material supports received from Special Study Group for Bioseparation Engineering, SCEJ, and Snow Brand Co. are gratefully acknowledged.

REFERENCES

1.P. Walstra, and R. Jenness, Dairy Chemistry and Physics, John Wiley & Sons, New York, 1984.

Bioseparation Engineering
I. Endo, T. Nagamune, S. Katoh and T. Yonemoto (Editors)

Separation of Dead Cells from Culture Broth by Using Dielectrophoresis

M. Hakoda and N. Shiragami

Department of Biological and Chemical Engineering, Gunma University, 1-5-1, Kiryu, Gunma 376-8515, Japan.

We apply the dielectrophoresis to separation of cells from culture broth by considering the difference of dielectric constant between alive and dead cells. In this paper, the dependency of the applied a-c voltage's frequency on the dielectrophoresis of cells is studied and the method of cell collection is also discussed. As the results, it is shown that the selective separation of alive cells is enable by dielectrophoresis. At that time, there was no detrimental effect even if applying the a-c electric field of 1000V/cm to alive cells.

1. INTRODUCTION

Many kinds of bioproduct are produced from biocatalysts such as microbial cells, animal cells and plant cells. In a large scale cultivation of these cells by a bioreactor, it is difficult for all cells to keep alive condition for a long time, and a part of cells are died. The technique for removing dead cells in a bioreactor is quite important in order to maintain the quality of product. We apply the dielectrophoresis to separation of cells from culture broth by considering the difference of dielectric constant between alive and dead cells.

In this paper, the dependency of the applied a-c voltage's frequency on the dielectrophoresis of cells is studied and the method of cell collection is also discussed. In addition, the influence of the applied a-c voltage on the activity of cells is examined.

2. THEORETICAL CONSIDERATION

Dielectrophoresis is defined as the motion of a neutral particle caused by polarization effects in a non-uniform electric field[1-2]. The non-uniform electric field is formed by applying a-c voltage between a pin electrode and a plate electrode. **Figure 1** illustrates the behaviors of particles in the non-uniform electric field. When the dielectric constant of particle is larger than that of medium, the particle moves towards the pin electrode of greatest electric field strength. Whereas the dielectric constant of particle is smaller than that of medium, the particle moves to the opposite direction.

The dielectrophoretic force $\mathbf{F}_e$ is represented as

$$\mathbf{F}_e = 2\pi r_p^3 \varepsilon_m \varepsilon_0 \left(\frac{\varepsilon_p - \varepsilon_m}{\varepsilon_p - 2\varepsilon_m} \right) \mathrm{grad} \, | \mathbf{E}_e |^2 \qquad (1)$$

where ε_m and ε_p are the absolute dielectric constants of the medium and that of the particle, respectively; ε_0 is the permittivity of vacuum space; $\mathbf{E}_e$ is the electric field strength; and r_p is the radius of particle.

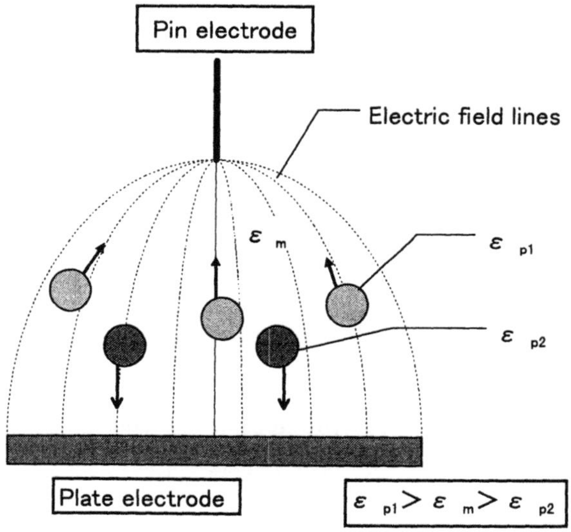

Fig.1 Comparison of behaviors of particle by difference of dielectric constant in a non-uniform electric field.

3. MATERIALS AND METHODS

In this experiment, yeast cells (*Saccharomyces cerevisiae*) were used. Yeast cells used were cultured on a solid agar plate in a culture dish (Iwaki Glass Co. Ltd.) at 30°C. The experimental apparatus is schematically shown in **Fig.2**. It consists of an dilelectrophoretic well, a microscope (BHC, Olympus Optical Co., Ltd.), a video camera (WV-CD20, National Co., Ltd.), a video monitor (WV-5410, National Co., Ltd.), a osillator (AG-203, Kenwood Co. Ltd.), a high speed power amplifier (4005, NF electronic Co. Ltd.) and a osilloscope (VP-526A, National Co. Ltd.). Suspension of cells were filled into a 2.4 ml well mounted pin-plate electrodes, and an a-c voltage was applied between electrodes. The suspension of cells was prepared by dispersing cultured yeast into distilled water. The ranges of the electric field strength and the frequency were from 2×10^3 V/m to 1×10^5 V/m and from 1kHz to 1MHz, respectively. The dead cells were obtained by autoclaving alive cells. The dielectrophoretic well is shown in **Fig.3**. The well and electrodes were made of acrylic resin plate and platinum wire, respectively. An aqueous suspension of yeast cells was placed in the well and an a-c voltage applied between the electrodes. Each experiment was carried out by using the alive or dead cells. All experiments were performed in a constant temperature room at 20°C.

Specific collection rate defined as Eq.(2) is used as a measure of the effectiveness of the collection.

$$\text{Specific collection rate} = \frac{\text{Number of collected cells}}{\text{Number of whole cells in well} \times \text{Time}} \qquad (2)$$

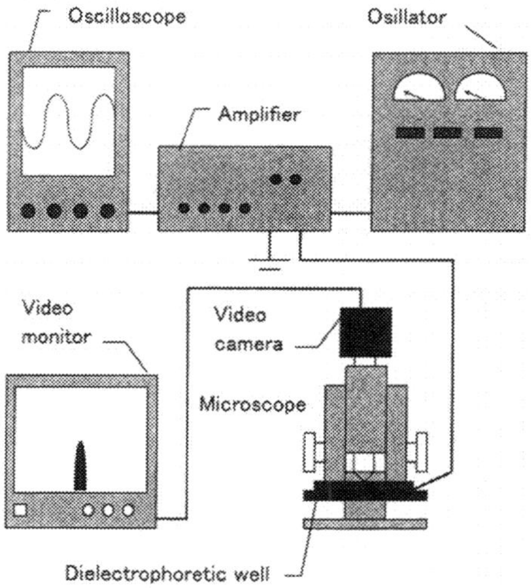

Fig.2 Experimental apparatus

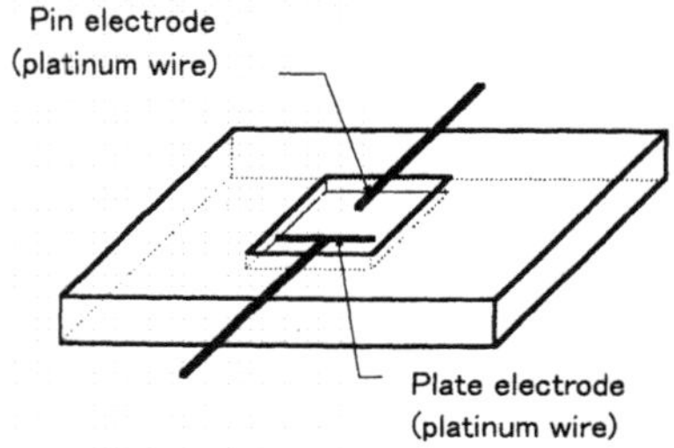

Fig.3 Schematic diagram of dielectrophoretic well

4. RESULTS AND DISCUSSION

A typical observation of collected cells on the pin electrode by the microscope is shown in **Fig.4**. It was found that the cells migrated towards the pin electrode and the number of collected cells increased gradually. The cells were collected onto the pin electrode as chains parallel to the electric field lines. The number of collected cells was counted microscopically for only tip region of the pin electrode .

tip of pin electrode

Fig.4 Typical result of collected cells on tip of pin electrode by microscope

4.1. Time course of collected cells

Dielectrophoresis for the alive cells was performed under the condition of 1MHz, 300 V/cm. **Figure 5** shows the time course of the number of the collected alive cells. The number of collected alive cells attains a saturated value within 60 min. At 60 min, the ratio of collected cells to whole cells in the well was about 4%. Since the number of collected cells was measured a narrow tip part of the pin electrode, the ratio was very small.

4.2. Effect of frequency

Each experiment with different frequency was carried out by using the alive or dead cells. The effect of the applied a-c voltage's frequency on the specific collection rate is shown in **Fig.6**. The number of collected alive cells strongly depended on the frequency. The dead cells also migrated towards the pin electrode, however, the dependency of the frequency was quite different from that

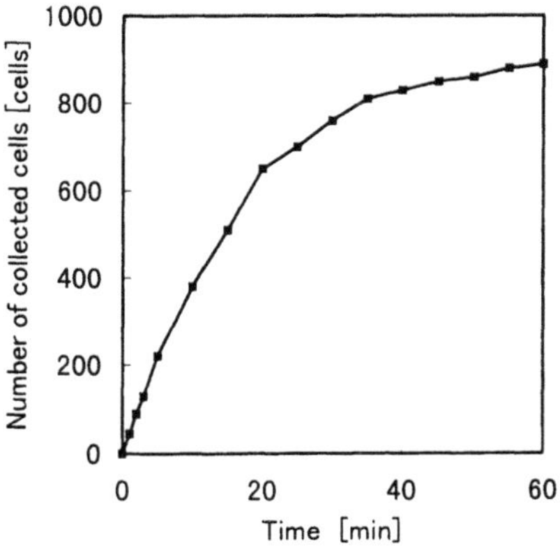

Fig. 5 Time course of number of collected cells. 1 MHz, 300V/cm.

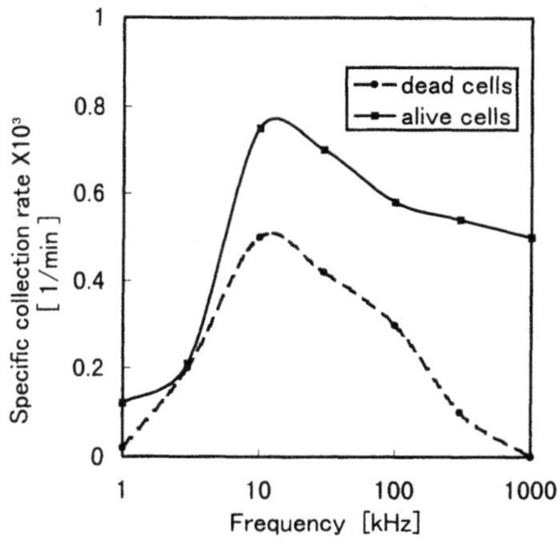

Fig. 6 Effect of frequency on specific collection rate. 300 V/cm, 1 min.

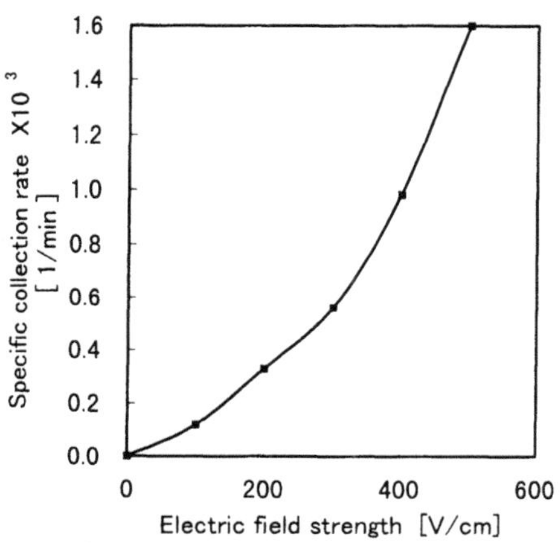

Fig. 7 Effect of electric field strength on specific collection rate. 1 MHz, 1 min.

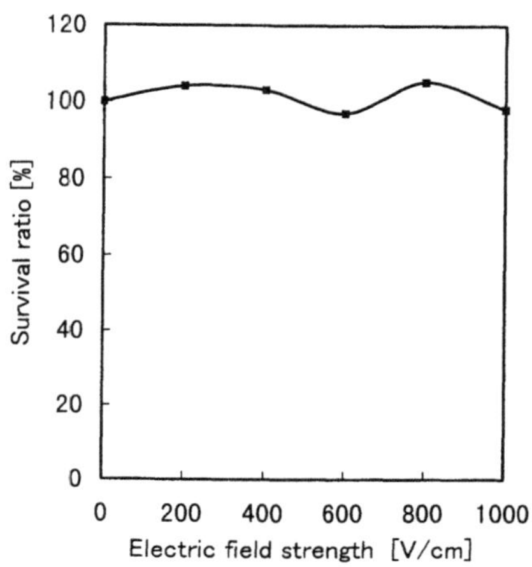

Fig. 8 Effect of electric field strength on survival ratio. 1 MHz, 30 min.

of alive cells. When the field was applied at lower frequency under 10 kHz, the osillated motion of cells was observed. The collected cells were removed from the pin electrode. Alive cells can be collected at the frequency of 1MHz, and dead cells never be collected at this frequency. From these results, it is shown that the selective separation of alive cells is enable by dielectrophoresis at high frequency.

4.3. Effect of electric field strength on specific collection rate

At the frequency of 1 MHz , the effect of electric field strength on the specific collection rate of alive cells is shown in **Fig.7**. The specific collection rate of alive cells in a given period increased with increasing the electric field strength. From the result, it is found that the higher electric field strength was effective.

4.4. Effect of electric field strength on survival ratio

If the collected cells were yet viable, there is possibility that the collection is performed at higher electric field strength. The influence of the electric field strength on a viability of cells was examined. The viability was evaluated by colony counting. The cells collecting by the pin electrode were placed on a solid agar plate and incubated for 2 days at 30°C. The number of colonies which grew from single alive cell was counted. **Figure 8** shows the effect of the electric field strength on a survival ratio of the collected cells at 1 MHz, 30 min. It is considered that there is no detrimental effect even if applying the a-c electric field strength of 1000V/cm to alive cells.

5. CONCLUSION

To enable a separation of dead cells from culture broth, an experiment by using dielectrophoresis was carried out. The number of collected alive cells in a given period increased with increasing the electric field strength, and strongly depended on the applied a-c voltage's frequency. Alive cells can be collected at the frequency of 1 MHz, and dead cells never be collected at this frequency. From these results, it is shown that the selective separation of alive cells is enable by dielectrophoresis. When the cells collecting by the pin electrode were cultured on a solid agar plate, these cells showed the normal growth. It is considered that there is no detrimental effect even if applying the a-c electric field strength of 1000V/cm to alive cells.

REFERENCES

1. H.A. Pohl, J. Appl. Phys., 22 (1951) 869
2. H.A. Pohl, Dielectrophoresis, Cambridge University Press, 1978

Bioseparation Engineering
I. Endo, T. Nagamune, S. Katoh and T. Yonemoto (Editors)

Microcalorimetric studies of interactions between proteins and hydrophobic ligands in hydrophobic interaction chromatography : effects of ligand chain length, density, and the amount of bound protein

Fu-Yung Lin[a], Wen-Yih Chen[a], Ruoh-Chyu Ruaan[b], and Hsiang-Ming Huang[a]

[a]Department of Chemical Engineering, National Central University Chung-Li, Taiwan 320
[b]Department of Chemical Engineering, Chung-Yuan University Chung-Li, Taiwan 320

Using Isothermal Titration Calorimetry (ITC), this investigation directly measured the adsorption enthalpies of proteins on various hydrophobic adsorbents. Various amounts of butyl and octyl groups were attached onto CM-Sepharose to form C4 and C8 two types of hydrophobic adsorbents. The adsorption enthalpies of both trypsinogen and α-chymotrypsinogen A were measured at 4.0 M NaCl and pH 10.0, in which most ionic interaction was suppressed. The adsorption isotherms of both proteins on various adsorbents were also measured, thus allowing us to calculate the Gibbs free energy and entropy of adsorption.

1. INTRODUCTION

Tiselius [1] first reported on HIC, observing the retention of dye in a so called salting-out chromatography where sulfate and phosphate solutions were present. HIC is an important method for separating and purifying biomaterials. The method is based on the interfacial interaction between biomaterial and hydrophobic ligands which are chemically attached to hydrophilic matrices. The studies on effects of the protein binding with hydrophobic adsorbents were widely investigated. However, studies adopting from the thermodynamic perspective are limited, and so far, the adsorption enthalpy of HIC has not been directly measured.

This study elucidates the effects on the binding mechanism by varying the ligand chain length, density of adsorbents. Also discussed herein are the differences in enthalpy change of two protein, α-chymotrypsinogen A and trypsinogen with two modified hydrophobic adsorbents, CM-butyl-Sepharose and CM-octyl-Sepharose under varying amounts of bound protein. By following the previous study [2-5] which demonstrated the binding mechanism of imidazole and lysozyme with immobilized Cu(II) and Fe(III) by varying pH values and NaCl concentrations in immobilized metal ion affinity chromatography (IMAC). The discussion of the binding mechanism in HIC can also be divided into five sequential subprocesses: (a) the dehydration of the protein; (b) the dehydration of the adsorbent ; (c) hydrophobic interaction between proteins and hydrophobic adsorbents; (d) the conformation of the protein; and (e) rearrangement of the excluded water or ion molecules in a bulk solution increasing entropy.

2. RESULTS AND DISCUSSION

2.1. Effects of ligand Chain Length

Figures 1 and 2 show the ΔH_{ads} values of both proteins with CM-octyl-Sepharose at constant ligand density were higher than with CM-butyl-Sepharose and the difference in value was 7 ~ 8 kJ/mol. Specifically, the dehydration of hydrophobic ligand of the adsorbent and the hydrophobic interaction between protein and hydrophobic ligands (i.e., (b) and (c) processes) should be considered mainly for the difference in this study. Therefore, the difference Figs. 1 and 2 shown in the value of ΔH_{ads} for CM-butyl-Sepharose and CM-octyl-Sepharose implies that the heat required for the dehydration (i.e., the (b) process) differs. The reported data [6] also supported this finding indicating that the heat required to dehydrate the n-octyl chain was 10 kJ/mol higher than that required to dehydrate the n-butyl chain. This result also agrees with the study of Horvath et al. [7]. Furthermore, the ΔH_{ads} value of CM-octyl-Sepharose was positive implying the adsorption for CM-octyl-Sepharose was mostly by entropy. Therefore, we further conclude that the interaction mechanism of both proteins with CM-octyl-Sepharose is more like partitioning, while with the CM-butyl-Sepharose it is an adsorption dominated process. This conclusion is similar to the summary of Dorsey and Dill [8].

2.2. Effects of Ligand Density

The adsorption enthalpy of both proteins increased with ligand density as shown in Figures 1 and 2. Possible explanations are (i) a reduction in the endothermic amount of the dehydration (i.e., (a) and (b) processes) at a lower ligand density ; (ii) an increase in the exothermic amount of the hydrophobic interaction (i.e., the (c) process) ; (iii) a reduction in the endothermic amount for structural rearrangement of the protein (i.e., the (d) process) at a higher ligand density of the adsorbents. We believe the major contributor to the result was the heat required for the dehydration. Furthermore, the adsorption entropies of both proteins increased with ligand density, as showed in Figs. 3 and 4, implying the contribution of the dehydration process to the entropy-gain increased with ligand density. This result agrees with the study of Cole and Dorsey [9]. They found that calculated by van't Hoff plots the enthalpy and entropy values of benzene with C18 ligands, both increased with ligand density.

2.3. Effects of the amount of bound protein

In this study, adsorption enthalpies (ΔH_{ads}) of both proteins increased with the amount of bound protein at a constant ligand chain length and density of the adsorbent at 4.0 M NaCl and pH 10.0 as Figs. 1 and 2 show. The results can be explained by (i) the protein-protein interaction being enhanced by an increase in the amount of bound protein ; (ii) the hydrophobic bonding of the protein with the adsorbent (i.e., the (c) process) minimized by an increase in the bound protein ; (iii) the heat required for the dehydration of the adsorbent (i.e. (a) and (b) processes) reducing as the amount of bound protein increases. As discussion above, we believe that protein-protein interaction was the major contributor to the results obtained.

3. CONCLUSIONS

The adsorption enthalpies of both proteins on octyl containing adsorbents were all positive, which indicated that the adsorption on C8 adsorbents were mostly entropy driven. However, the adsorption enthalpies on butyl containing adsorbents were all negative. On low butyl content, 23 μmol/mL gel, the adsorption enthalpies contributed more than one third of the negativity of the adsorption Gibbs free energy. Therefore, the adsorption on low density C4 adsorbents should have been partially enthalpy driven. As the butyl density increased, the adsorption enthalpy became more positive. The adsorption process became entropy driven again. The above results indicate that increase resin hydrophobicity, by increasing either alkyl length or density, increased the adsorption enthalpy and entropy. Thus the positivity of adsorption enthalpy, directly measurable by Isothermal Titration Calorimetry, apparently could serve as an index for hydrophobic interaction.

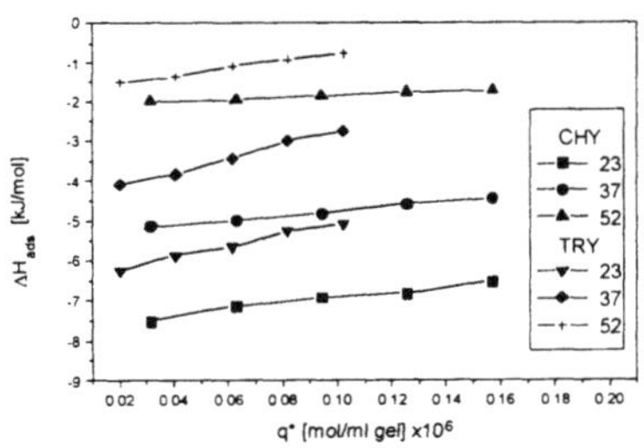

Figures 1. Adsorption enthalpies (ΔH_{ads}) of α-chymotrypsinogen A or trypsinogen adsorption onto CM-butyl-Sepharose with variable ligand density at 4.0 M NaCl and pH 10.0.

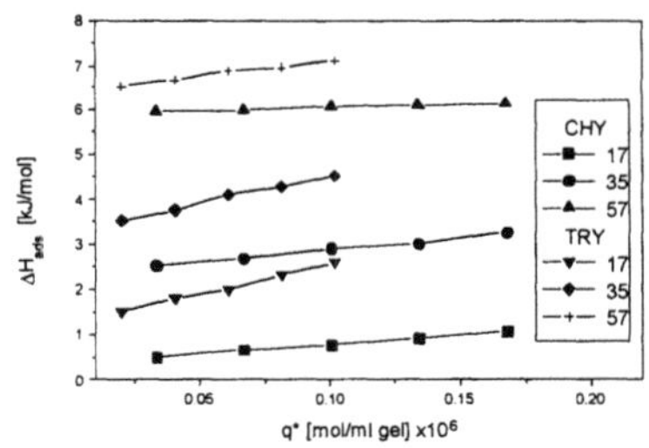

Figures 2. Adsorption enthalpies (ΔH_{ads}) of α-chymotrypsinogen A or trypsinogen adsorption onto CM-octyl-Sepharose with variable ligand density at 4.0 M NaCl and pH 10.0.

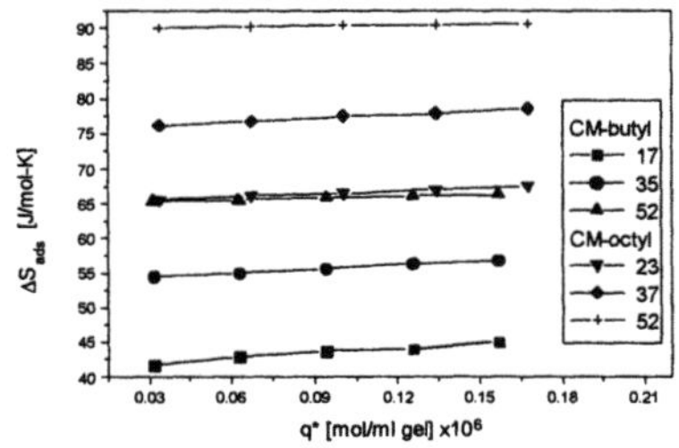

Figures 3. Adsorption entropies (ΔS_{ads}) of α-chymotrypsinogen A adsorption onto CM-butyl-Sepharose or CM-octyl-Sepharose with variable ligand density at 4.0 M NaCl and pH 10.0.

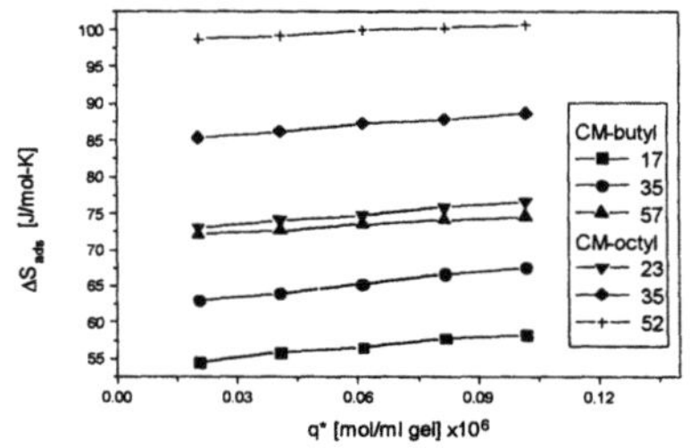

Figures 4. Adsorption entropies (ΔS_{ads}) of trypsinogen adsorption onto CM-butyl-Sepharose or CM-octyl-Sepharose with variable ligand density at 4.0 M NaCl and pH 10.0.

REFERENCES

1. C.C. Shepard, A. Tiselius, in: Chromatographic Analysis, Discussions of the Faraday Society; Hazell Watson and Winey, Ltd, London, 1949; No. 7, p.275.
2. C.F. Wu, W.Y. Chen and J.F. Lee, J. Colloid Interface Sci. 28 (1995) 419.
3. W.Y. Chen, J.F. Lee, C.F. Wu and H.K. Tsao, J. Colloid Interface Sci. 190 (1997) 649.
4. F.Y. Lin, W.Y. Chen and L.C. Sang, J. Colloid Interface Sci. 214 (1999) 373.
5. W.Y. Chen, F.Y. Lin and C.F. Wu, in: J. Stubenrauch, (Ed.), Interface Dynamic, Marcel Dekker, New York, 1999, in publication.
6. I. Makhatafze and P.L. Privalov, J. Mol. Biol. 232 (1993) 639.
7. D. Haidacher, A. Vailaya and Cs. Horvath, Proc. Natl. Acad. USA. 93, (1996) 2290.
8. J.G. Dorsey and K.A. Dill, Chem. Rev. 89 (1989) 331.
9. L.A. Cole, and J.G. Dorsey, Anal. Chem. 64 (1992) 1317.

Bioseparation Engineering
I. Endo, T. Nagamune, S. Katoh and T. Yonemoto (Editors)

Membrane Phase Separation of Aqueous/Alcohol Biphase Mixture and Its Application for Enzyme Bioreactor.

Y. Isono[a,b] and M. Nakajima[a*]

[a]National Food Research Institute, Ministry of Agriculture, Forestry and Fisheries, 2-1-2, Kannondai, Tsukuba, Ibaraki 305-8642, Japan

[b]Dainichiseika Color & Chemicals Mfg. Co., Ltd., 1-9-4, Horinouchi, Adachi-ku, Tokyo 123-8555, Japan

An availability of membranes for a phase separation of an aqueous/alcohol mixture was examined. The single alcohol phase could be obtained as the permeate through a hydrophobic microfiltration membrane without visible leak of the aqueous phase. The flux of 1-hexanol was the highest among seven different alcohols. The separability and alcohol flux were affected by both of the alcohol solubility in water and the alcohol viscosity. Continuous membrane phase separation of 1-hexanol from the aqueous/1-hexanol mixture could be performed stably, and in which the average flux was 10×10^{-6} $m^3 \cdot m^{-2} \cdot s^{-1}$. Similarly, the single aqueous phase could be obtained as the permeate through a hydrophilic ultrafiltration membrane without visible leak of the alcohol phase. Continuous membrane phase separation of the aqueous phase from the aqueous/1-hexanol mixture was also performed, in which the average flux was 5.4×10^{-6} $m^3 \cdot m^{-2} \cdot s^{-1}$. Furthermore, the membrane phase separation technique was applied to an enzyme bioreactor using 1-hexanol/water biphasic reaction system to synthesis an aspartame precursor (ZAPM). According to the partition coefficient, ZAPM was extracted into the 1-hexanol phase. Then, 1-hexanol phase containing ZAPM was obtained as permeate through the hydrophobic membrane. Integration of membrane phase separation into the biphasic enzyme reaction system could be realized stable ZAPM production continuously.

1. INTRODUCTION

Phase separation of an aqueous/ organic solvents mixture is an important process in many industrial fields. Removing oil from wastewater is a notable aspect in an environmental field. Phase separation on the end of biphasic extraction is also crucial in chemical, biological, and environmental processes. Recently, membrane phase separation technique has received much attention in the recent decade. Separation of an oil phase from oily wastewater using a membrane has been actively investigated for the purpose of oil removal from waste water.[1, 2] The membrane filtration of emulsion for demulsification in an extraction process[3] and a combined system of an extraction and a membrane separation for bioreactor system[4,5] were also studied.

*corresponding author. Tel : +81-298-38-8025 Fax : 81-298-38-8212 e-mail : mnaka@nfri.affrc.go.jp

Biphase extraction technique has been widely applied to a bioreactor system. A reaction equilibrium can be improved by removing the product in the reaction phase using an extraction. We have demonstrated a perstraction technique, which is organic solvent extraction through a membrane, and its application to a bioreactor system for synthesis of the aspartame precursor(ZAPM).[6)] In the perstraction system, the streams of aqueous and organic phases are separated by a membrane, i.e. the phase dispersion and separation can be eliminated. Although high purity of the product was obtained with the system, mass transfer rate of solute was insufficient due to molecular diffusion through the membrane. The integration of biphasic extraction with membrane separation of biphase for bioreactor system would be favorable to improve the mass transfer rate and the productivity of the reactor system. However, few investigations have been performed in this field.

The present research was undertaken to confirm the availability of a membrane for the phase separation of aqueous/alcohol biphase mixture which has been reported as suitable for the extractive bioreactor system for the synthesis of the aspartame precursor previously.[7)] To apply the membrane phase separation into the bioreactor, a preliminary study of ZAPM synthesis using extraction-membrane phase separation system was also performed.

2. EXPERIMENTAL

2.1. Materials

Water immiscible alcohols, 1-butanol, 2-butanol, 2-metyl-1-propanol (iso-butanol), 1-heptanol, 1-hexanol, 1-pentanol, and 2-metyl-2-propanol (*tert*-pentanol) were purchased from Wako Pure Chemical Industries, Ltd., Japan. McIlvaine buffer (5×10^{-2} mol · dm^{-3}) saturated with an alcohol and an alcohol saturated with McIlvaine buffer were used for the aqueous and the alcohol phases, respectively. The hydrophobic membrane (PE; material, polyethylene; pore size, 0.03 μm) and hydrophilic membrane (NTU-4208; material, polyimide; molecular weight cut off, 8000) were supplied from Tonen Chemical Co., Ltd. and Nitto Denko Corporation, respectively.

2.2. Filtration procedures

The membrane was attached in a membrane filtration cell (membrane diameter, 4.3×10^{-2} m; effective membrane area, 1.45×10^{-3} m^2; cell volume, 2.0×10^{-4} m^3), and was washed with a permeation phase prior to the experiments. Then, the adequate volume ratio of the aqueous and alcohol phases were supplied to the cell, and mixed with a stirring bar at 40℃. The biphase mixture was filtered by supplying pressure with nitrogen gas, and the permeate flux was measured. The permeate flux at continuous filtration was measured in the same way as that of batch filtration except that the permeation phase was continuously supplied to the cell from the stainless steel reservoir using pressurized nitrogen gas.

2.3. Enzymatic synthesis of ZAPM

Both the aqueous phase, McIlvaine buffer (5×10^{-2} mol · dm^{-3}) saturated with hexanol at pH 6.0 containing 20 g · dm^{-3} of the enzyme, and the organic phase, hexanol saturated with McIlvaine buffer containing 15 mmol · dm^{-3} of ZA and 100 mmol · dm^{-3} of PM, were prepared. Initially, 0.05 dm^3 of aqueous phase and 0.05 dm^3 of organic phase were fed into the cell. Continuous ZAPM synthesis was performed in a manner similar to the measurement of the solvent flux at continuous filtration except that the organic phase

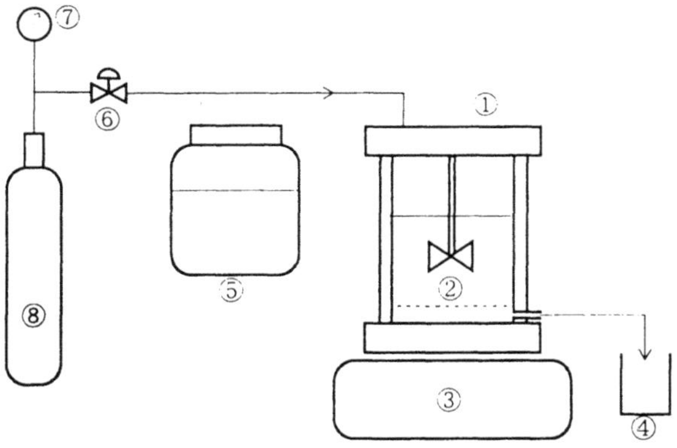

Fig. 1 Filtration apparatus
① membrane filtration cell
② membrane
③ magnetic stirrer
④ permeate reservoir
⑤ stainless steel reservoir
⑥ pressure controller
⑦ pressure gauge
⑧ nitrogen gas cylinder

containing the substrates. The average retention time of the organic phase was controlled at 160 min. The concentration changes of substances in the permeate were measured using HPLC at regular time intervals.

3. RESULTS AND DISCUSSION

3.1. Alcohol Phase Permeation

Fig. 2 shows the effect of the alcohol phase fraction on the flux of 1-hexanol using PE membrane. The alcohol phase was dispersed in continuous aqueous phase. The single alcohol phase could be obtained without visible leak of aqueous phase in which alcohol phase fraction ranged from 0.03 to 0.3. The flux decreased with decreasing of the alcohol phase fraction, probably due to that the smaller fraction has lower probability of contacting with the membrane. No permeation of both phases was observed with the membranes at the alcohol phase fraction of 0.03. Shrinkage of membrane was not observed in PE membrane after the experiment, therefore it is recognized that the membrane can be stable for 1-hexanol filtration.

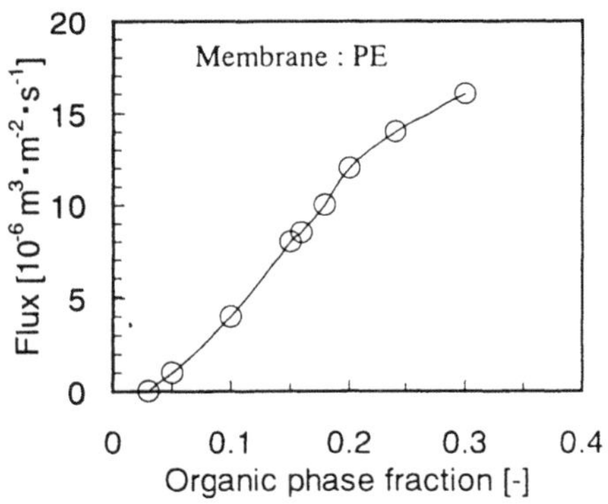

Fig. 2 Effect of the alcohol phase fraction on the 1- hexanol flux

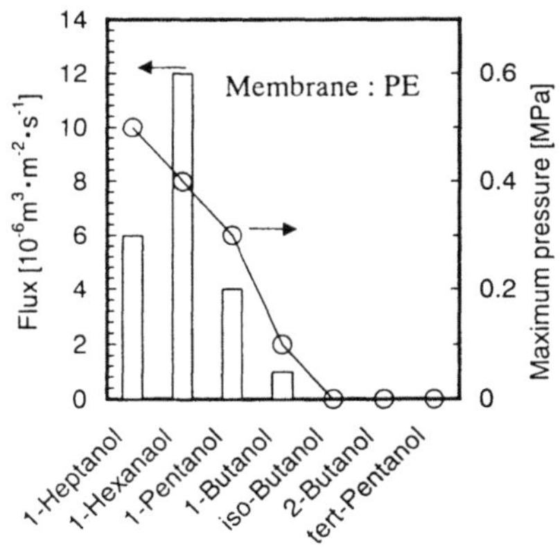

Fig. 3 Separation results with various alcohols

Fig. 3 shows the separation results with alcohols when alcohol fraction was 0.2. The alcohol phases of 2-butanol, iso-butanol and *tert*-pentanol did not permeate through the membrane at the operating pressure of 0.1 MPa, and both alcohol and aqueous phases permeated at the pressure higher than 0.1 MPa. These alcohols have higher solubility in water (more than 8.0%), then the alcohol phase could not be separated from the mixture at any operating pressure. It is difficult to separate the alcohol phase from the mixture when the alcohol solubility in water is high, because high alcohol solubility may increase the affinity between the two phases. Four alcohols of 1-butanol, 1-heptanol, 1-hexanol, and 1-pentanol could be separated from the biphase mixture as the permeate without visible leak of aqueous phase. Fig. 4 shows the relationship between the alcohol viscosity and the permeability of single phase alcohols from the biphase mixture. The available pressure, which is defined as the pressure to obtain the alcohol phase with the highest flux without visible leak of aqueous phase (shown in Fig. 3), decreased according to increase of alcohol solubility in water. Vaidya *et al.*[8] presented that the available pressure is directly proportional to the liquid-liquid interfacial tension. A mutual solubility between alcohol and water is related to the interfacial tension. Therefore, the available pressure may depend on the alcohol solubility. Although the available pressure of 1-hexanol was lower than that of 1-heptanol, 1-hexanol showed the highest flux. The viscosity of 1-heptanol is higher than that of 1-hexanol, therefore, 1-hexanol showed lower flux than that of 1-hexanol. In these results, both of the alcohol solubility and the alcohol viscosity affect the separability and alcohol flux. It was revealed that 1-hexanol is suitable for the separation system.

Fig. 5 shows the 1-hexanol permeation flux in the continuous separation at 0.3 MPa operating pressure from the mixture in which alcohol phase fraction is 0.2. The alcohol phase could be obtained continuously as permeate without visible leak of aqueous phase. Stable flux was obtained during the continuous operation, and the average flux of 1-hexanol was about 10 $\times 10^{-6}$ $m^3 \cdot m^{-2} \cdot s^{-1}$, which is similar to that with batch operation.

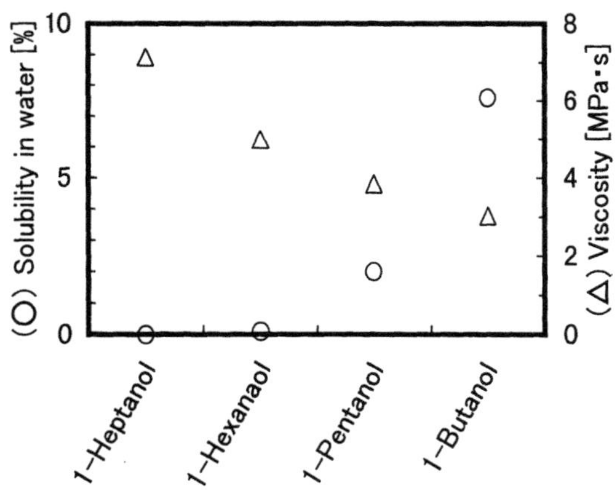

Fig. 4 Alcohol solubility in water and the permeability.

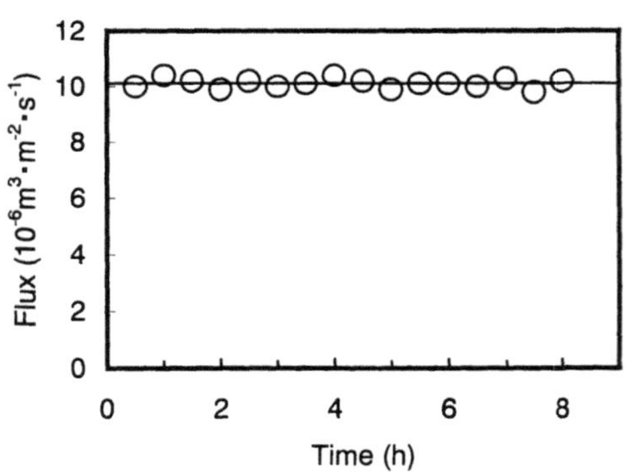

Fig. 5 Flux of 1-hexanol in the continuous separation with PE membrane.

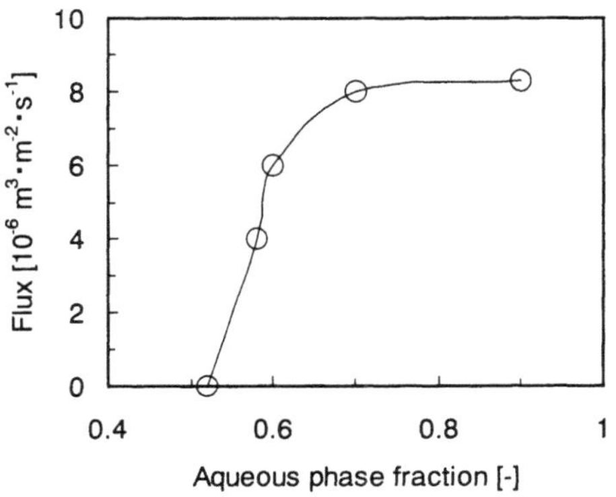

Fig. 6 Effect of the aqueous phase fraction on the aqueous phase flux.

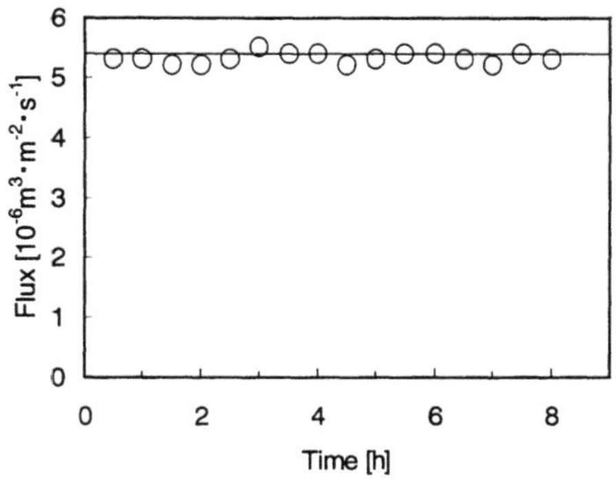

Fig. 7 Flux of the aqueous phase in the continuous separation with NTU-4208 membrane.

3.2. Aqueous Phase Permeation

Fig. 6 shows the effect of the aqueous phase fraction on the aqueous phase flux with NTU-4208 membrane. The alcohol phase was 1-hexanol, and the operating pressure was 0.15 MPa. The aqueous phase permeation without visible leak of aqueous phase could be performed with the membranes in the aqueous phase fraction ranged from 0.52 to 0.9. The flux decreased drastically with decreasing of the aqueous phase fraction, of which the reason is possibly similar that in the alcohol phase permeation.

Fig. 7 shows the aqueous phase flux in the continuous separation at 0.15 MPa operating pressure from the mixture in which aqueous phase fraction is 0.6. The single aqueous phase could be obtained continuously as permeate without visible leak of alcohol phase. Stable flux was obtained during the continuous operation, and the average flux was about $5.4\times10^{-6}\ m^3\cdot m^{-2}\cdot s^{-1}$, which is similar to that with batch operation.

3.3. ZAPM synthesis

The time courses of ZAPM synthesis in the aqueous and organic phases by the reactor system are illustrated in Figs. 8(a) and 8(b), respectively. The visible leak of aqueous phase was not observed during the 26 h for which the experiment was performed. The stable permeate flux which was calculated to be $4.2\times10^{-6}\ m^3\cdot m^{-2}\cdot s^{-1}$ could be obtained during the reaction period. Stable ZAPM synthesis was performed continuously by the enzymatic reaction, and the concentration of ZAPM in the permeate organic phase was kept higher than that in the aqueous phase according to the partition coefficient. This result revealed that the ZAPM synthesized in the aqueous phase by enzymatic reaction was extracted into the organic phase, and then the organic phase containing ZAPM was separated from the biphase mixture by the membrane as the permeate. The present system can realize the recovery of the product from the reaction media using biphasic extraction and the separation of the extraction phase using membrane phase separation simultaneously. The ZAPM productivity per volume of

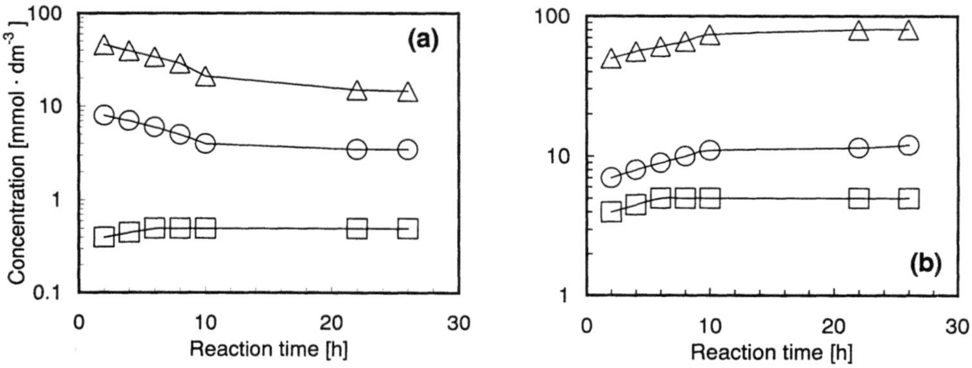

Fig. 8 Time courses of ZAPM synthesis in the aqueous phase(a) and organic phase(b) by the enzyme reactor system. Symbols; ○, ZA; △, PM; □, ZAPM.

both phases was determined to be 8.4 kg · m^{-3} · d^{-1}. Comparison of this result to the productivity data (6.6 kg · m^{-3} · d^{-1}) in our previous report [10] shows that an improvement in the bioreactor for ZAPM synthesis has been achieved. Additionally, no enzyme leakage observed during the operation period in our system due to the use the membrane. Our system seems to overcome a common problem in establishing a simple mixer-settler system is how to avoid enzyme entrainment. The proposed combination system of biphasic extraction of the bioproduct and the separation of the organic phase after the extraction step using hydrophobic membranes is an effective method for the ZAPM production.

REFERENCES

1) G. K. Anderson and C. B. Saw, Environ. Technol. Lett., 8(1987)121.
2) P. A. Bailey, Filtr. Sep., 14(1997)53.
3) K. Scott, I. F. McConvey and A Adhamy, J. Membr. Sci., 72(1992)245.
4) W. Pronk, M. van der Burgt, G. Boswinkel, and K. van't Riet, J. Am. Oil Chem. Soc., 68(1991)852.
5) Y. Sahashi, and H. Ishizuka, Kagaku Kogaku Ronbunshu, 20(1994)148.
6) Y. Isono, K. Fukushima, G. Araya, H. Nabetani and M. Nakajima, J. Chem. Technol. Biotechnol., 70(1997) 171.
7) Y. Isono, H. Nabetani and M. Nakajima, Nippon Shokuhin Kagaku Kogaku Kaishi, 42(1995)920.
8) A. M. Vaidya, G. Bell and P. J. Halling, J. Membr. Sci., 71(1992)139.
9) J. D'Ans and E. Lax, Taschenbuch fur Chemikeru. Physiker, 2nd ed., Springer, p.959, 1949., cited in Kagaku Kogaku Kyokai (Eds.), Kagaku Kogaku Binran, Maruzen, Tokyo, p.83. 1988.
10) Y. Isono, H. Nabetani and M. Nakajima, Process Biochem., 30(1995) 773.

Bioseparation Engineering
I. Endo, T. Nagamune, S. Katoh and T. Yonemoto (Editors)

Microfabricated Structures for Bioseparation

Jong Wook Hong[a], Kazuo Hosokawa[b], Teruo Fujii[c], Minoru Seki[a], and Isao Endo[d]

[a]Department of Chemistry and Biotechnology, Graduate School of Engineering, The University of Tokyo, Tokyo 113-8656, Japan
[b]Surface and Interface Technology Division, Mechanical Engineering Laboratory, Ibaraki 305-8564, Japan
[c]Institute of Industrial Science, The University of Tokyo, Tokyo 106-8558, Japan
[d]Biochemical Systems Laboratory, The Institute of Physical & Chemical Research (RIKEN), Saitama 351-0198, Japan

As a new apparatus for bioseparation, a PDMS (polydimethylsiloxane) microchip for capillary gel electrophoresis that can separate different sizes of biomolecules in a small and simple experimental scale is presented. The chip was made by microfabrication technology and the micro-capillary structures on the microchip were partially filled with agarose gel that can enhance separation resolution. Biomolecules such as DNA molecules were successfully driven by electric field and separated to form bands in the agarose-filled microstructures.
Keywords: Microfabrication, Bioseparation, PDMS (polydimethylsiloxane), Biomolecules

1. INTRODUCTION

Separation and purification of biochemicals is difficult and very often costs more than the production process of them. Since large-scale production of bioproducts had been started, processes such as filtration, extraction, adsorption, chromatography and electrophoresis have been studied and used for the separation and purification of large amount of bioproducts from microbial, plant or animal cell culture. On the other hand, with the emerging of the new field named "μ(micro)-Biochemical Engineering" or "Micro-biosystems", equipments for the separation of biochemicals in volume of micro to nano liter scale are required and studied [1]. In this paper, a microchip having microfabricated structure on it for the separation of biomolecules is presented along with the advantages of micro-scale separation as well as microfabrication technology.

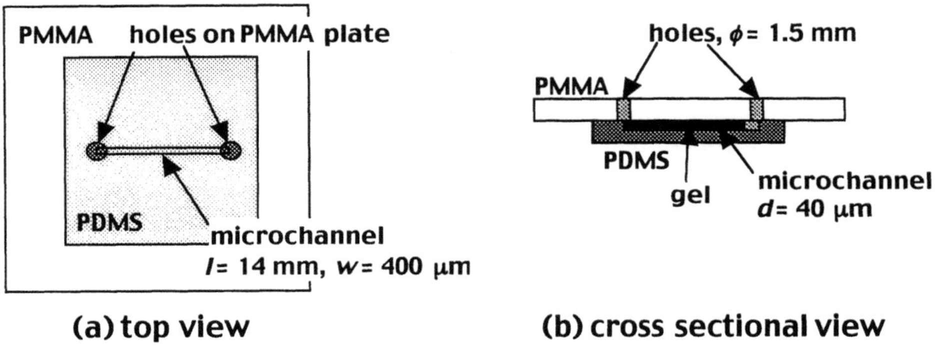

Fig. 1. Layout of the typical PDMS microchip used for bioseparation (not to scale). (a) Top view of the chip and (b) Cross sectional view of PDMS microchip attached under PMMA plate.

2. MICROFABRICATION

Microfabrication comprises the use of a set of manufacturing tools based on photolithographic techniques commonly used for making IC or LSI in the electronics industry. Microfabrication is also common key technology in the promising new fields such as MEMS (microelectro-mechannical systems), μTAS (micro total analytical systems), and μFLUMS (micro fluidics and molecules) and has been applied to immunoisolation, drug delivery, biochemical engineering and analytical fields including capillary electrophoresis (CE) [2-6]. The potential benefits of micro-sized apparatus, relative to systems of conventional size, include reduced consumption of samples and reagents, shorter analysis times, greater sensitivity, portability that allows in situ and real-time analysis and disposability [7].

3. MATERIALS AND METHODS

3.1. Fabrication of master

The length of micro-structured channel is 14 mm long. The width is from 30 μm to 800 μm and the depth is controlled from 10 μm to 150 μm. Microchannel patterns were printed in high resolution (4,064 dpi) on transparencies, which were used as masks in photolithography. Ultrathick photoresist (SU-8; Microlithography Chemicals Co., MA) is spin-coated onto silicon wafer to create masters with micro structures. The fabrication process of the PDMS microchip for the separation of biomolecules is outlined in Fig. 2. A silicon wafer was dried for dehydra-

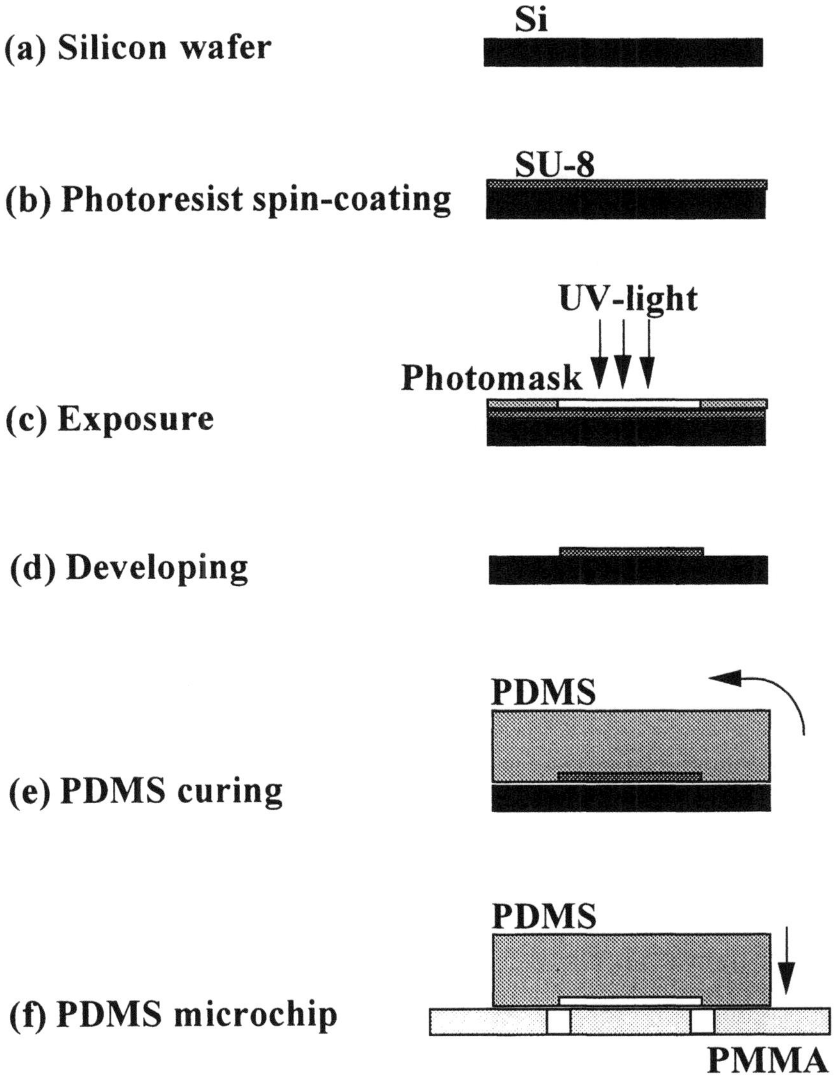

Fig. 2. Fabrication process of the PDMS chip. (a) A silicon wafer was cut to 20 mm x 20 mm. (b) Ultrathick negative photoresist SU-8 was spin-coated on the silicon wafer. (c) Pholithographic polymerization of negative photoresist. After this procedure, the silicon wafer was developed . (d) Prepolymer of PDMS was poured onto the master and solidified . (e) The cured PDMS was peeled off from the master. (f) The PDMS chip was pasted on a PMMA plate

tion and spin-coated, and baked in an oven. This procedure was repeated to create feature of photoresist so that the microstructures from 10 μm to 150 μm were realized. By using a mask aligner (PEM-800; Union Optical Co., Tokyo, Japan), the pattern on the mask was pholitographically transferred to the photoresist coated silicon wafer. After development in (1-Methosy-2-propyl)-acetate (Merck Co.), the master was washed in isopropylalcohol and distilled water. Before pouring prepolymer of PDMS, by using RIE (Reactive Ion Etching System 10NR; Samco Co., Japan), the master was treated with fluorocarbon (CHF_3) to facilitates the easy removal of the PDMS replica after molding.

3.2. Curing of PDMS and gel preparation

Prepolymer of PDMS and curing agent (Sylgard 184; Dow Corning Co., MI) was mixed with the ratio from 5:1 to 10:1, stirred thoroughly and then degassed in vacuum. The prepolymer mixture was poured onto the master and cured at the range of 65°C to 150°C. After curing, the PDMS replica was peeled off from the master. This PDMS replica was attached onto a PMMA (polymethylmethacrylate) substrate having two holes which serve as the ports. The PDMS microchip with PMMA substrate was oxidized by O_2 plasma for surface hydrophilization. Agarose powder was dissolved in 1 x TBE (Tris borate EDTA) buffer using a microwave oven. This agarose solution was kept at 65°C in oven and carefully introduced into the micro channel on the PDMS microchip by capillary action and matured at room temperature.

3.3. Sample, sample loading and separation

Two mL of FITC (fluoresceinisothiocyanate) labeled 100 bp DNA ladder solution was placed in the sample loading port and the potential was applied between port 1 and 2 through Pt electrodes connected to a power supply. The sample DNA was concentrated and loaded at the edge of the agarose gel by electric force. After sample was loaded to the gel in the fabricated channel structures, the solution in a sample loading port was replaced by 1 x TBE buffer and separation was carried out applying electrical voltage through the channel.

3.4. Detection and data analysis

Fluorescence of FITC labeled DNA was detected by an inverted fluorescence microscope (DIAPHOT-TMD; Nikon Co., Japan) an ICCD camera (C2400-80; Hamamatsu Photonics Co., Japan). The image of electrophoresis was recorded with 8 mm video camera. After the experiments, the recorded image of electrophoresis was digitized by an image analysis program, NIH image.

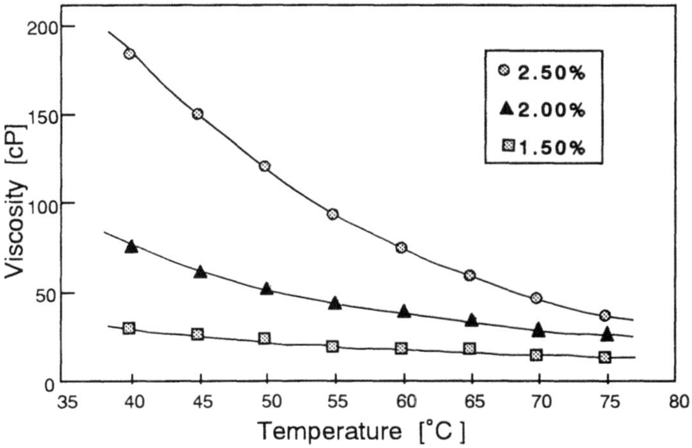

Fig. 3. Viscosity of agarose solutions at various temperatures. SeaKem GTG agarpse powder was dissolved in 1 X TBE buffer solution.

3. RESULTS AND DISCUSSION

The preparation of gel-filled capillaries of uniform quality and stability has proven difficult in general. Gel instability during electrophoretical separation, *i.e.* bubble formation and clogging of pores in the gels, limits the field strength and the number of effectively possible runs in a gel [8]. For the evaluation of the best conditions for gel preparation in the PDMS microchip, the relationship between temperatures and viscosity was revealed under the temperature condition ranging from 40°C to 75°C (Fig. 3). Although a trial to load 2.5% agarose solution to the microchannel was carried out at 70°C and 75°C, the agarose solution did not move into the microchannel because of the high viscosity. On the other hand, agarose solution of 2.0% could be introduced to the microchannel above 65°C at which the viscosity was 33 cp. Therefore, the separation was performed in 2.0% agarose in TBE buffer. In our experiments, the agarose was introduced into the surface modified microchannel and gelated within 10 min for the entire process.

Applying 100 V for 1 sec through the gel filled microchannel successfully formed sample plug. In the PDMS microchip-based bioseparation system, the sample amount could be drastically reduced to less than several nano-liter. The field strength for the separation is under about 70 V/ cm. In the microfabricated PDMS chip, separation of DNA molecules ranging from 100 bp to 1 kbp was completed within 8 mm in 120 sec which is several to 20 fold faster than conventional capillary or slab gel electrophoresis.

REFERENCES

1. M. Seki, J.W. Hong, T. Fujii and I. Endo, Proc. of the Annual Meeting of the Young Asian Biochemical Engineers Community (1999) 13-15.
2. M. Madow, Fundamentals of Microfabrication, CRC Press, New York, 1997.
3. NSF, Engineering Microsystems: XYZ on a Chip, http://www.nsf.gov/getpub?nsf9931.
4. T.A. Densai, W.H. Chu, J.K. Tu, G.M. Beattle, K.A. Hayek, and M. Ferrai, Biotech. Bioeng., 57 (1998) 118-120.
5. J.T. Santini Jr., M.J. Cima and R. Langer, Science, 397 (1999) 335-337.
6. J.W. Hong, K. Hosokawa, T. Fujii, M. Seki and I. Endo, IEEE Transducers'99 Digest of Technical Papers 1 (1999) 760-763.
7. A. Manz, D.J. Harrison, E.M.J. Verpoorte, J.C. Fettinger, A. Paulus, H. Luedi and M. Widmer, J. Chromatography, 593 (1992) 253-258.
8. W.M. Sunada and H.W. Blanch, Electrophoresis, 18 (1997) 2243-2254.

Bioseparation Engineering
I. Endo, T. Nagamune, S. Katoh and T. Yonemoto (Editors)

Production of a human IgM-type antibody and preparation of combinatorial library by recombinant *Saccharomyces cerevisiae*

N. SHIOMI[a], K. MURAO[b], H. KOGA[b] and S. KATOH[c]

[a]Department of Human Sciences, Kobe College, 1-4 Nishinomiya,651-8505, Japan

[b]Department of Chemical Science and Engineering, Kobe University, Nada Rokkodai, Kobe, 657-8501, Japan

[c]Graduate School of Science and Technology, Kobe University, Nada Rokkodai, Kobe, 657-8501, Japan

Anti-exotoxin A human antibody genes (IgM) were cloned from the mouse-human hybridoma and expressed in *S. cerevisiae*. The recombinant *S. cerevisiae* cells having these vectors could secret the anti-exotoxin A antibody into culture broth. By using mating characteristics in *S. cerevisiae*, we suggested a simple method for preparation of a huge-scale library of antibodies.

1. INTRODUCTION

Human antibodies which can bind to proteins of cancer cell, virus, toxin etc. are useful for therapy. Mouse and human-mouse chimera antibodies, however, have not been used for these purposes, because these antibodies had antigenicity at the parts derived from mouse. Therefore, production methods for human antibodies have been studied actively.

Expression of genetic engineered library of human antibodies is one of the important method for production of human antibodies, and has recently been investigated by using phage, *Escherichia coli* and animal cells as host cells.[1)] Expression of a library by using yeast has not been reported, although yeast cells had been used as convenient hosts for expression of proteins derived from human cells. It was reported that recombinant *Saccharomyces cerevisiae* could secret human IgG antibody into culture broth[2)] but failed to secret human IgM antibody.[3)] In this paper, we investigated the production of human IgM type antibody against exotoxin A from *Pseudomonas aeruginosa* by recombinant *S. cerevisiae* cells. Furthermore, we proposed a simple method for preparation of huge-scale libraries of antibodies in *S. cerevisiae*.

2. MATERIALS AND METHODS

2.1 Genetic method

S. cerevisaie ATCC52053 (*MATa met2-1 ura3-52 leu2-3,112, lys2*), ATCC60729 (*MATα MAL6 mal1 ura3-52 leu2-3,112 trp1 his*) and ATCC60730 (*MAT*a *MAL6 mal1 ura3-52 leu2-3,112 trp1 his*) were used as hosts, and SD medium (2% glucose, 0.67% yeast nitrogen base w/o amino acids) was used for the growth of *S.cerevisae*. *E. coli* DH5 and plasmid YEp24, pJDB207, pT7 Blue, pUC19 and pGLDP31-RcT were used for construction of expression vectors.

2.2 Purification by affinity chromatography

Anti-human IgM antibody was coupled to CNBr-activated Sepharose 4B and the immunoabsorbent was packed into a column of 1cm i.d. The culture broth was applied to the column at a flow rate of 1 ml/min. After washing with PBS, adsorbed IgM-type antibody was eluted with 0.1N HCl.

2.3 ELISA

Microtiter plate with 96 flat-bottomed wells was used for ELISA. Direct and sandwich ELISA were applied to assay the concentration of anti-IgM antibody and coupling with exotoxin A. In the direct ELISA, each of the wells was coated with sample solutions, washing with PBS, and blocked with diluted Block Ace. After washing with PBS containing 0.05 vol% Tween 20 (PBST), each well was incubated with anti-human IgM (rabbit) horseradish peroxide solution. Finally color was developed using 2,2-azino-di(3-ethylbenzthiazoline) sulfonic acid (ABTS) solution, and the absorbance of each well was measured by a microplate reader at 405nm. In sandwich ELISA, each well was firstly coated with exotoxin A, washing with PBS and blocked with diluted Block Ace. After washing, samples were incubated, and the coupled IgM was detected by the same method in direct ELISA.

3. RESULTS

3.1 Construction of the expression vectors of human IgM type antibody

Firstly, we constructed the vectors for expression of the genes of light and heavy chains of human IgM. The promoter gene of glycerolaldehyde-3-phospate dehydrogenase (GLDp) and the terminator gene of posphoglycerol kinase (PGK) were amplified by PCR method with a plasmid pGLD P31-RcT as a template DNA. These DNAs were digested with *Bam*HI and *Xho*I, and inserted into the T7 Blue vector. On the other hand, the signal gene of α-factor was also amplified by PCR method and inserted into the T7 Blue vector. The signal gene of α-factor was digested from the

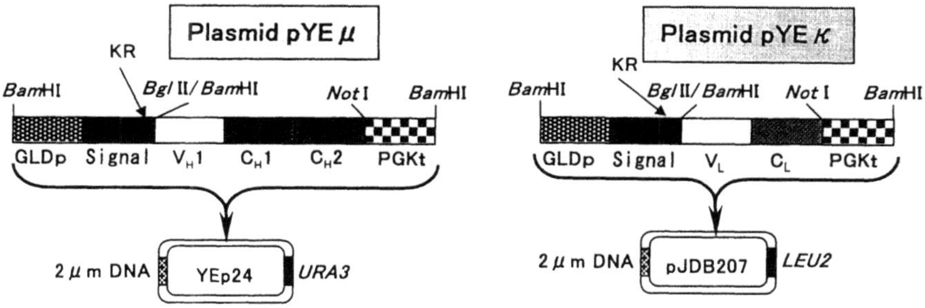

Figure1. Expression vectors of human antibody genes

plasmid, and inserted between GLDp and PGKt genes. The DNA fragment (1.5 kb), which contained GLDp, signal and PGKt genes, was digested with *Bam*HI and inserted into the *Bam*HI sites of YEp24 and pJDB207. Thus, two vectors for the expression of light and heavy chains were finally constructed (Fig.1) and named pAB1 and pAB2.

The human-mouse hybridoma ATCC HB8789, which could secret human antibody (IgM type) coupling with exotoxin A from *P. aeruginosa*, was used as a model antibody. After growth of hybridoma cells, mRNAs were extracted and cDNAs were synthesized with these mRNAs. Whole antibody genes were cloned by PCR method and nucleotide sequences were determined (data not shown). Then, VH1, CH1 and CH2 domains of μ gene (shown by μ *) and κ gene were amplified by PCR method with primers, and amplified DNA fragments were digested with *Bam*HI and *Not*I. The DNA fragments containing κ and μ * genes were inserted to the plasmids pAB1 and pAB2 digested with *Bgl*II and *Not*I. Finally, the vectors pYE μ and pYE κ were constructed, as shown in Fig. 1. To secret the antibody having correct N-terminal sequences, the antibody genes were inserted after Lys-Arg sequence of signal peptide, which was endopeptidase processing site.

3.2 Production of IgM-type antibody gene by *S. cerevisaie*

The plasmids pYE μ and pYE κ, which contained μ *and κ genes, were transformed into *S. cerevisiae* ATCC60730 and ATCC60729 cells, respectively. Both the plasmids were also co-transformed into ATCC52053 cells. The transformants containing pYE μ, pYE κ and the both plasmids were named ATCC60730/ μ, ATCC60729/ κ and ATCC52053/ μ κ, respectively. These transformants were cultured until stationary phase in SD medium for 3 days at 30°C. The culture broth was purified by anti-human IgM antibody-Sepharose 4B column.

The eluted fractions and culture broths of *S. cerevisiae* ATCC60730/ μ and ATCC60729/ κ cells were assayed by direct ELISA (Fig. 2, 3). Those of *S. cerevisiae* ATCC60730/ μ and ATCC60729/ κ cells reacted with anti-human IgM antibodies.

Therefore, we confirmed that *S. cerevisiae* ATCC60730/μ and ATCC60729/κ cells could secret heavy (μ*) and light (κ) chain in culture broth. Secondary, the elution and broth of ATCC52053/μ κ cells were examined by direct and sandwich ELISA. Those of ATCC52053/μ κ cells reacted with anti-human IgM antibody (Fig. 4) and exotoxin A (Fig. 5). These results suggested that *S. cerevisiae* ATCC52053 cells having both μ and κ genes (ATCC52053/μ κ) could secret the human IgM antibody into broth, and the secreted antibody with light and heavy chains could recognize exotoxin A .

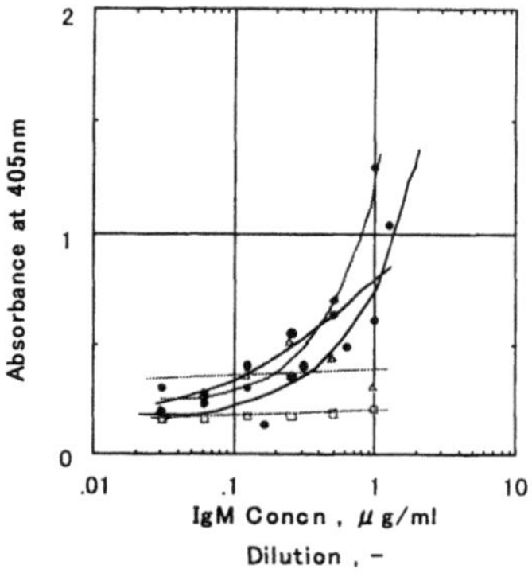

Figure 2. Secretion of heavy chain from *S. cerevisiae* 60730

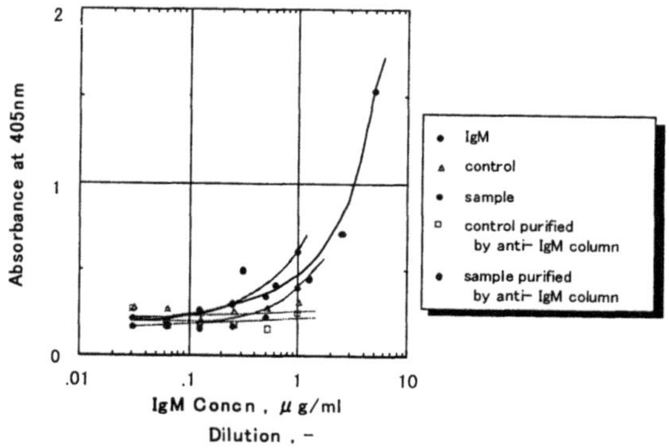

Figure 3. Secretion of light chain from *S. cerevivisa* 60729

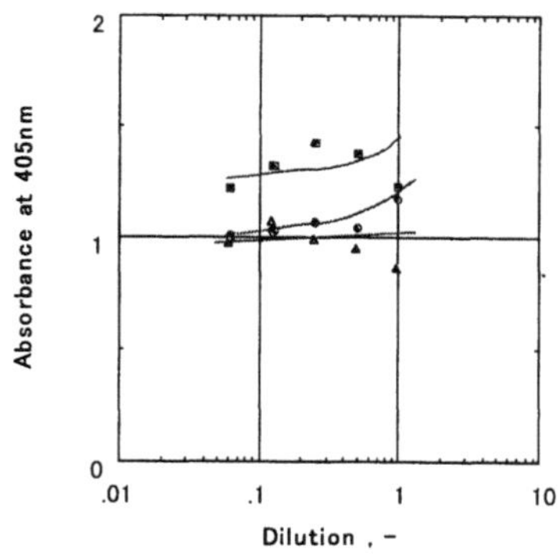

Figure 4. Screretion of light and heavy chain from *S. cerevisiae* 52053 (direct ELISA)

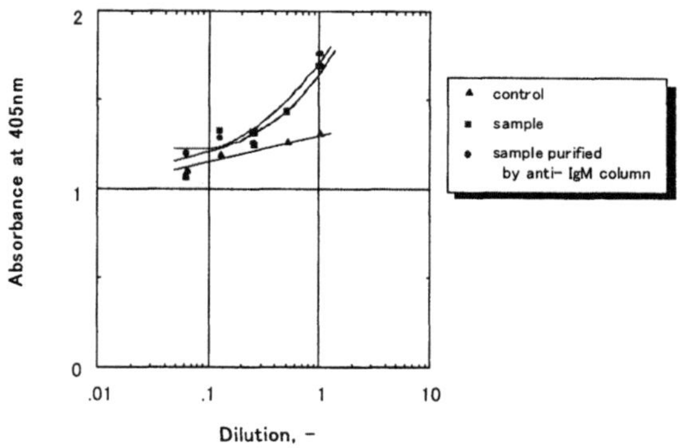

Figure 5. Screretion of light and heavy chain from *S. cerevisiae* 52053
(sandwich ELISA with exotoxin A)

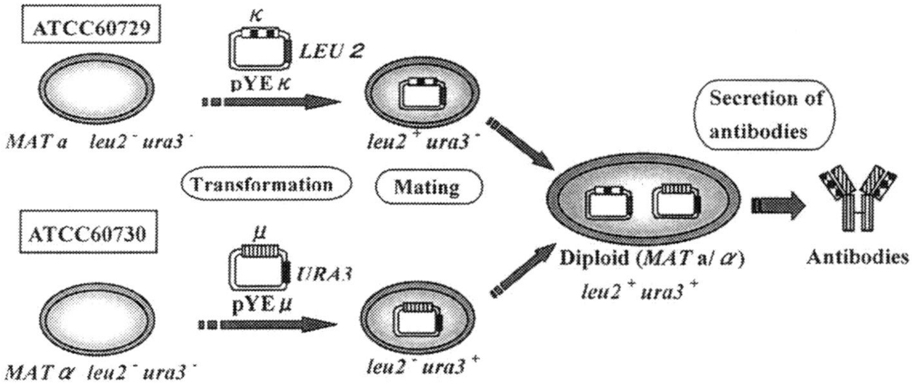

Figure 6. Production of human antibodies by mating a and α types of yeast cells

4. DISCUSSION

From the results in this work, yeast *S. cerevisiae* cells could secrete IgM-type antibody in culture broth (Figs. 2-5). These results suggest that genetic engineered library of human antibodies could be prepared in *S. cerevisiae* cells as a host. Thus, we tried to develop a simple method for large-scale library preparation of human antibody by use of characteristics of mating of yeast cells. The scheme of our proposed method is shown in Fig. 5. Two expression vectors containing μ * and κ gene are transformed into a and α types strains in *S. cerevisiae* ATCC60730 and ATCC60729, respectively. After these transformants are cultured, these cells are mixed each other and cultured overnight. By this mating operation, a/α type diploid cells, which have both pYEμ and pYEκ plasmids and secrete antibodies, can easily be prepared.

In the genetic engineered combinatorial library, huge-scale (10^8-10^{10} kinds) of library seems to be required for the isolation of useful antibodies that have high coupling constant. But it was hard to isolate 10^8-10^{10} kinds of the transformants in usual method. In our method, 10^{10} kinds of transformants were easily prepared by mixing 10^5 kinds of a and α types of transformants. It is suggested that our method of library preparation greatly useful for production of a huge-scale combinatorial library, because high efficiency in transformation is not required. We are now investigating characteristics of antibodies secreted from this diploid cells, and the results will be reported in another paper.

REFERENCES

1) R. J. Owens and R. J. Young, J. Immuno. Methods, 168 (1994) 149.
2) A. H. Horwitz *et al.*, Proc. Natl. Acad. USA, 85 (1988) 8678.
3) C. R. Wood *et al.*, Nature, 314 (1985) 446.

Bioseparation Engineering
I. Endo, T. Nagamune, S. Katoh and T. Yonemoto (Editors)

Dynamic binding performance of large biomolecules such as γ-globulin, viruses and virus-like particles on various chromatographic supports

Shuichi Yamamoto* and Eiji Miyagawa

Department of Chemical Engineering, Yamaguchi University
Tokiwadai, Ube 755-8611, Japan
*E.mail shu-yama@po.cc.yamaguchi-u.ac.jp fax +81-836-35-9933

Dynamic binding (adsorption) performance of γ-globulin, several viruses and virus-like particles was examined with various types of ion-exchange chromatography (IEC) supports. Among various IEC supports, a sulfated cellulose bead adsorbent (Cellufine-Sulfate) was carefully investigated as this is known to show unique bioaffinity and size exclusion behavior. Size exclusion curves and HETP-flow rate curves under non-binding conditions showed that proteins and viruses are totally excluded from Cellufine-Sulfate beads, and they can adsorb only on the surface layer of the bead. Biospecificity to Cellufine-Sultate was tested with various proteins, viruses and virus-like particles. Dynamic adsorption capacities (DBC) were determined as a function of flow-rate. DBC of γ-globulin on Cellufine-Sulfate did not depend on the flow velocity whereas conventional porous agarose-based IEC supports show a very strong flow-velocity dependent DBC.

1. INTRODUCTION

Ion exchange chromatography (IEC) is an efficient method for separating small and medium size proteins (molecular weight up to 70,000), and widely used not only in laboratory but also in biotechnology production processes[1-6]. IEC is also used for the separation of much larger biomolecules such as immunoglobulins, plasmid DNAs and viruses[7-9]. In contrast to large number of publications on protein IEC, research on separation of such large biomolecules by IEC is still not adequate.

In this paper, dynamic binding (adsorption) performance of γ-globulin, several viruses and virus-like particle was examined with various types of IEC supports. Among various IEC supports, a sulfated cellulose bead adsorbent (Cellufine-Sulfate) was carefully investigated as this is known to show rather unique bioaffinity, and size exclusion behavior[10,11]. For example, Human hepatitis B virus surface antigen (HBsAg) particles were efficiently separated on Cellufine-Sulfate whereas they were not adsorbed on conventional porous agarose-based IEC supports[11]. Size exclusion curves and HETP-flow rate curves were prepared from isocratic elution experimental data under non-binding conditions. Biospecificity to Cellufine Sulfate was tested with various proteins, viruses and virus-like particle. Dynamic adsorption capacities were determined as a function of flow-rate.

2.EXPERIMENTAL

2.1 Virus preparations and their assays

Human hepatitis B virus surface antigen (HBsAg) particles were purified on Cellufine-Sulfate chromatography, anti-HBs IgG Sepharose immunoaffinity and Sepharose CL-6B from the culture fluid of PLC/PRF/5 human liver carcinoma cell line (the Alexander cell) [12]. The purified HBV preparation had more than 95% purity by SDS polyacrylamide gel electrophoresis. HBsAg was assayed by a commercial enzyme immuno assay kit (FRELISA II Hbs Ag, Fujirebio, Tokyo, Japan). HBsAg concentration was determined from the absorbance at 280 nm (A280 was assumed to 10 for 10mg/mL)

Human parvovirus B19 (HPV B19) positive plasma with the infectivity titer of 10^6 $TCID_{50}$ /ml was used for the chromatography. HPV B19 was monitored by the infectivity assay using KU812Ep6 cell line[13].

2.2 Chromatography supports, protein and other solute, and equipment

Cellufine Sulfate m (Sulfate Cellulofine m)(Chisso, Tokyo, Japan) is a spherical cellulose bead (particle diameter 44-105 μm) having a sulfate ester group (8μM/mL)[10]. Its partial structure is shown in Fig.1. SP-Sepharose FF ion-exchange supports (gels) (Amersham Pharmacia Biotech, Uppsala, Sweden) are 6% cross-linked spherical agarose gel beads having a sulphopropyl group (SP-) [4]. These beads were packed into plastic columns (ID 9 mm, packed bed height Z 15cm) for size-exclusion and adsorption studies (ID 9mm Z=5cm for SP-FF). Vitamin B12(MW=1356), standard globular proteins from gel filtration calibration kit (Amresham-Pharmacia Biotech) and acetone were used for the size exclusion experiment under non-binding conditions (10mM phosphate pH8 containing 0.5 M NaCl). Bovine serum albumin, human-γ-globulin, lysozyme, ribonuclease A, and urease were purchased from Sigma (USA). Vitamin B12 and other reagents were obtained from Wako Pure Chemical (Osaka, Japan).

A fully automated liquid chromatography (LC) system (Prosys workstation, Beckman, Fullerton, USA) was used for the size exclusion and the adsorption studies with standard proteins. A simple LC setup (Pharmacia) with an open column (ID 16mm Z =2cm) was used for viruses and HBsAg particle separation(adsorption) experiments. The column experiments were done at 298 ±1K. The details of chromatographic conditions are shown in the figure captions.

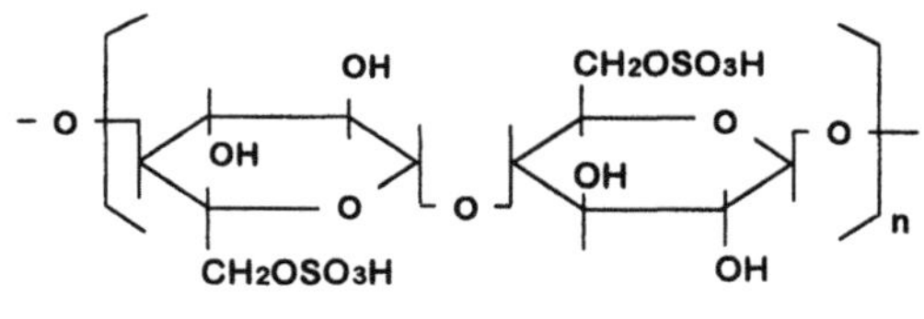

Figure 1 Structure of sulfated cellulose beads, Cellufine-Sulfate[10].

3. RESULTS AND DISCUSSION

It is well known that slow mass transfer of large biomolecules such as proteins in pores of packing media (supports or beads) governs adsorption performance in chromatography[3]. This is more critical when very large proteins or viruses and virus-like particles are to be separated. Cross-linked 6%-agarose gel beads (Sepharose) are commonly used for protein separation. The pore size is large enough for most proteins [1,3,4,14]. However, the dynamic adsorption capacity (DBC) of proteins is known to decrease drastically with an increase in molecular weight from *ca.*70000 (albumin) to 160000 (globulin) [1,4,5]. This indicates that very slow mass transfer (diffusion) rates in the pores govern the dynamic adsorption capacity even with supports having large pores. As HBsAg particles are much larger than most proteins(*ca.* 20nm in diameter[15,16]), the diffusion rate in the pore of the agarose bead becomes very low . We calculated the molecular diffusion coefficient D_m of HBsAg at 298K with its diameter as *ca.* 2×10^{-11} m^2/s by the equation proposed by Tyn and Gusek[17]. When the obstruction factor in the pore $\gamma_s = D_s/D_m$ is assumed to be $\gamma_s = K/4 = 0.1/4 = 0.025$ [3,18], the pore diffusivity D_s becomes *ca.* 5×10^{-13} m^2/s (The distribution coefficient K was estimated as *ca* 0.1 from our size exclusion curve measurements. See Fig.2). This values is much lower than an estimated D_s value (1×10^{-11} m^2/s) for a medium size protein such as bovine serum albumin (mol. wt. 68000) .

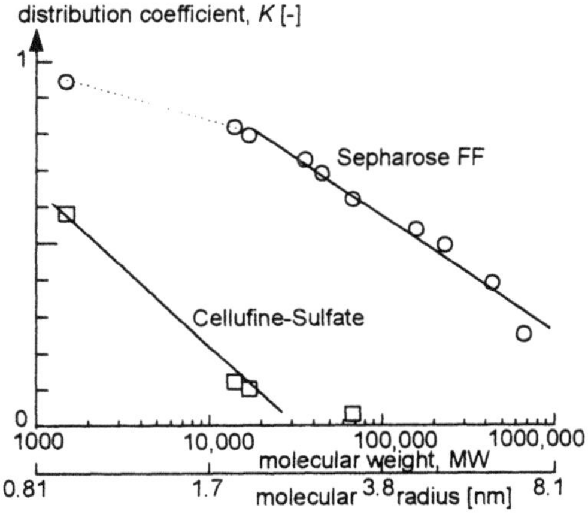

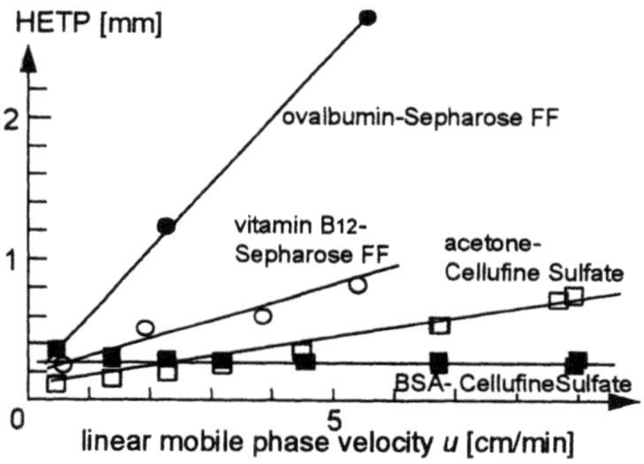

Figure 2(left). Size exclusion curves (distribution coefficient K *vs*. molecular weight MW) at non-binding conditions. K was determined from the retention volume V_R as $K=(V_R-V_0)/(V_t-V_0)$ where V_t = total bed volume, V_0 = void volume determined from the V_R of Dextran T2000. This K is equivalent to K_{av} in the literature [1]. Mobile phase: 10 mM phosphate buffer (pH 8) containing 0.5 M NaCl as the mobile phase (isocratic). column size: 0.9cm ID, 15cm length (Cellufine-Sulfate); 0.9cm ID, 15cm length (SP-Sepharose FF). The molecular radius was calculated by $0.081MW^{1/3}$ [18]. The data for acetone were not shown in the figure.

Figure 3(right). HETP as a funcion of linear velocity. HETP = $Z(w_v/V_R)^2/8$ where Z=column (bed) length, w_v= peak width in mL at $C=0.3679C_{max}$ (C_{max}=maximum peak height, C=peak height). The other conditions are the same in Fig.2.

The pore size of Cellufine Sulfate beads is so small that proteins are excluded and only very small molecules are able to diffuse into the pores [10-11]. This was verified by the size exclusion curve under non-binding conditions (Fig.2). The distribution coefficient (K) value of acetone was 0.65, which indicates the bead is highly crosslinked so that even a very small molecule can not utilize the intraparticle (pore) volume fully. Even a small protein like ribonucelase (mol.wt. 18000) was almost excluded as K<0.08. Bovine serum albumin (BSA, mol.wt. 68000) was completely excluded (K<0.02). When the molecule is excluded, the plate height (HETP) is known to be independent of the flow-rate[3,13]. As shown in Fig.3 the HETP values of BSA are constant and the values of acetone increase with the flow velocity. These findings indicate that proteins and other larger molecules are adsorbed only on the surface of Cellufine Sulfate beads under binding (adsorption) conditions at relatively high adsorption/desorption rates as pore diffusion is not involved.

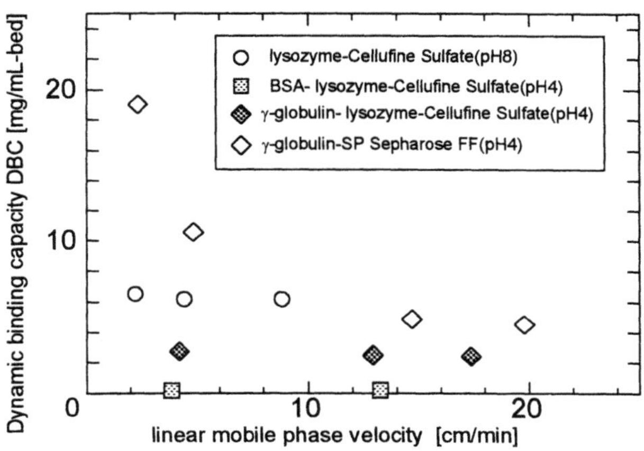

Figure 4 Dynamic binding capacity (DBC) as a function of linear mobile phase velocity. DBC=$C_0(V_B-V_0)/V_t$ where V_B = breakthrough volume at the relative concentration = 10% . Mobile phase = 10 mM phosphate buffer containing 0.03M NaCl (pH8) or 20 mM sodium acetate buffer containing 0.03M NaCl (pH4). Column (packed bed) size : 0.9cm ID, 15cm length (Cellufine-Sulfate); 0.9cm ID, 5cm length (SP-Sepharose FF). Note that the data point DBC = 47 mg-BSA/mL-SP Sepharose FF(pH4) at 20 cm/min is not included in the figure.

As we did not have enough amounts of HBsAg particles for dynamic binding capacity measurements, γ-globulin (mol.wt. 150,000) was chosen as a model protein. Lysozyme and bovine serum albumin were also used for comparison. As shown in Fig.4, the dynamic binding capacity (DBC) of lysozyme (*ca.*7 mg/mL-gel) and of γ-globulin (*ca.* 3 mg/mL-gel) on Cellufine Sulfate did not depend on the flow velocity whereas the DBC of γ-globulin for SP-Sepharose (porous agarose-based IEC support) decreased markedly with an increase in flow velocity because of slow diffusion in the pores [3,20,21]. As proteins adsorb only onto the surface of Cellufine Sulfate, the adsorption rate is high although the equilibrium capacity is much lower than conventional porous agarose ion-exchange beads (for example, 100-150 mg-lysozyme/mL-gel, Ref.1,4). Because of this rapid adsorption rate, the DBC values of γ-globulin for Cellufine Sulfate at high flow rate regions are similar to those for SP-Sepharose FF. Advantages and disadvantages of the use of the surface layer of ion-exchange supports were already discussed in detail [22,23]. Even with a macroporous resin initial adsorption rates are high as they are controlled by mass transfer to the outermost subparticles.

Bovine serum albumin (BSA) was not adsorbed onto Cellufine Sulfate As this can not be explained by a simple electrostatic interaction mechanism, molecular recognition of Cellufine Sulfate might be different from agarose-based IEC supports. Adsorption of various proteins and viruses onto Cellufine Sulfate was tested (Table 1). Polysaccharide sulfate (heparinoid) is known to have some specific affinity to such biomolecules as lipoproteins and some viruses. This may explain the biological recognition mechanism of sulfated-cellulose beads summarized in Table 1.

Table 1 Summary for adsorption on Cellufine Sulfate (sulfated cellulose) adsorbents

solute	adsorption[1)]	desorption	adsorption capacity[2)]
γ-Globulin	Yes	0.5M NaCl	2.5-7.5 mg/mL-gel[3)]
Bovine serum albumin	No		
Lysozyme	Yes	0.5M NaCl	7 mg/mL-gel
Urease	No		
Human Hepatitis B surface antigen (HBsAg)	Yes	3M NaCl	20μg/mL-gel[4)]
Human parvovirus B19(HPVB19)	No		
Human immunodeficiency virus type 1 (HIV-1)	Yes	3M NaCl	ND
Human T cell leukemia virus type I (HTLV-I)	Yes	3M NaCl	ND
Human adenovirus type 7	No		
Baculovirus	Yes	3M NaCl	ND

1)Proteins were charged onto the column equilibrated with 10mM phosphate buffer (pH 8.0) containing 0.03 M NaCl while viruses were charged onto the column equilibrated with 10mM Tris-HCl buffer (pH 8.0) containing 0.15 M NaCl. Adsorption tests were carried out with packed beds. Namely, test solutions were charged onto the column, and the effluent was monitored.

2) DBC= dynamic binding capacity = C_0V_B/V_t where C_0=initial concentration, V_B=effluent volume at 10% breakthrough and V_t=total bed volume.

3) DBC values measured as a function of pH (4.0 to 6.0) varied from 2.5 to 7.5 mg/mL-gel.

4) This is not the precise DBC value but to show an approximate value of adsorption capacity.

5) Not determined.

In conclusion, only the sulfate groups on the surface of Cellufine -Sulfate bead (cellulose bead having a sulfate ether group) is accessible for large molecules (mol.wt.>70000). This is unfavorable in terms of the total adsorption capacity but is favorable for relatively rapid adsorption/desorption rates of large molecules such as viruses and virus-like particles. There is a biological recognition (biospecific affinity) mechanism between polysaccharide sulfate (heparinoid) and some biological products, which may also play an important role in retention of virus and virus-like particles.

Acknowledgment
This work was supported by a Grant-in Aid for scientific research on Priority Areas(No.296) (Grant No.11132255) and (C2, No.10650746) from the Ministry of Education, Science, Sports and Culture, Japan

REFERENCE

1. E.Karlsson, L.Ryden and J.Brewer, Ion-exchange chromatography, in Protein purification 2nd ed., p.145, ed. by J-C.Janson & L.Ryden, Wiley-VCH, New York, 1998
2. G.Sofer and L.Hagel, Handbook of Process Chromatography, Academic Press, San Diego, 1997
3. S.Yamamoto, K.Nakanishi and R.Matsuno, Ion-exchange chromatography of proteins, Marcel Dekker, New York, 1988
4. Ion-exchange chromatography Principles and Methods 3rd ed, Pharmacia Biotech, 1991
5. R.K.Scopes, Protein purification 2nd ed., Springer-Verlag, New York, 1987
6. F.Regnier, Methods in Enzymol., 104, (1984) 170
7. F.Blanche, Downstream, p.16, vol.27, Amersham Pharmacia Biotech, 1998
8. Anonymous, Downstream, p.125, vol.27, Amersham Pharmacia Biotech, 1998
9. A.Foriers, B.Rombaut and A.Boeye, J.Chromatogra., 498 (1990) 105
10. Sulfate-Cellulofine Application notes, Chisso, Tokyo, Japan, 1989
11. S.Yamamoto and E.Miyagawa, J.Chromatogra.A., 852(1999)25
12. P.L.Marion, F.H.Salazar, J.J.Alexander and W.S.Robinson, J.Virology, 32(1979)796
13. E.Miyagawa, T.Yoshida, H.Takahashi, K.Yamaguchi,T.Nagano,Y.Kiriyama,K.Okochi and H.Sato, " Infection of human parvovirus B19 to erythroid cell line, KU812Ep6 and application to titration of B19 infectivity", J. Virological Methods, in press
14. J-C.Janson and J-A.Jonsson, Introduction to chromatography, in Protein purification 2nd ed., pp.43-78, ed. by J-C.Janson and L.Ryden, Wiley-VCH, New York, 1998
15. K.Koike, E.Yoshida, K.Katagiri, M.Katayanagi, M.Oda, H.Tsunoo, K.Yaginuma and M.Kobayashi, Jpn.J.Cancer Res. (Gann), 78 (1987) 1341
16. M.Belew, M.Yafang, L.Bin, J.Berglof and J-C.Janson, Bioseparation, 1(1991)397
17. M.T.Tyn and T.W.Gusek, Biotech.Bioeng., 35 (1990)327
18. L.Hagel, Gel Filtration, , in Protein purification 2nd ed., pp.79-143, ed. by J-C.Janson and L.Ryden, Wiley-VCH, New York, 1998
19. S.Yamamoto, M. Nomura and Y.Sano, J.Chromatogr., 394 (1987) 363.
20. L.E.Weaver and G.Carta, Biotechnol.Prog., 12 (1996) 342
21. H.A.Chase, J.Chromatogra., 297 (1984) 179
22. L.R.Snyder and J.J.Kirkland, Introduction to modern liquid chromatography, Wiley, New york, 1974
23. C.Horvath, Pellicular ion exchange resins in chromatography, in Ion exchange and solvent extraction, vol.5., pp.207-260, 1973

Bioseparation Engineering
I. Endo, T. Nagamune, S. Katoh and T. Yonemoto (Editors)

Effects of swelling pressure of resin and complex formation with a counter-ion on the apparent distribution coefficient of a saccharide onto a cation-exchange resin

S. Adachi and R. Matsuno

Division of Applied Life Sciences, Graduate School of Agriculture, Kyoto University, Sakyo-ku, Kyoto 606-8502, Japan

An equation for the apparent distribution coefficient K_{app} of a saccharide onto a cation-exchange resin was proposed, considering the swelling pressure of the resin and the complex formation of the saccharide and a counter-ion. The K_{app} values of glucose, mannose and fructose onto the resin in Ca^{2+} form were observed over their wide concentration ranges. The K_{app} of glucose became large as the solute concentration increased, while the K_{app} of fructose did small at the high concentrations. The K_{app} of mannose scarcely depended on its concentration. These could be explained by the equation.

1. INTRODUCTION

In designing a chromatographic separation process, the distribution coefficient of a solute onto a resin or gel is an important parameter to predict the elution time of the solute. The distribution coefficient of a saccharide has been treated as a constant that is independent of the solute concentration and of the existence of other components. However, such a treatment is doubtful from the following aspects: Ligand exchange, that is, complex formation between a solute and a counter-ion, has been believed to be the reason why electrically neutral saccharides with the same molecular mass can be separated using a cation-exchange resin [1]. If so, the coefficient seems to depend on the solute concentration. When a dense solution is applied to a bed packed with the resin, the resin shrinks, that is, the swelling pressure of the resin decreases. However, the effect of the decrease in swelling pressure on the distribution coefficient has not fully estimated.

Taking the effects of both the swelling pressure and the complex formation of the solute and the counter-ion into consideration, we have proposed an equation for the apparent distribution coefficient K_{app} [2].

In this paper, we will observe the K_{app} values of some hexoses onto a cation-exchange resin in Ca^{2+} form at various solute concentrations, and will discuss the results based on the proposed equation.

2. THEORETICAL

Figure 1 schematically shows a model for explaining the K_{app} of a solute onto a cation-exchange resin. The solute S partitions to the resin phase with an intrinsic distribution coefficient K, and forms a complex with the counter-ion M^+ with a binding constant B.

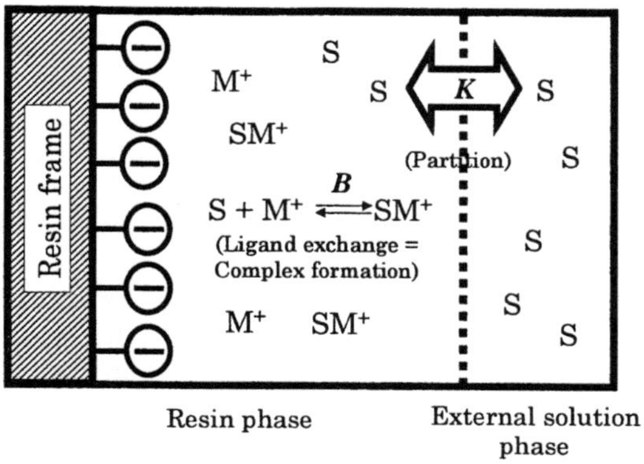

Figure 1. Illustration of a model for explaining the apparent distribution coefficient of a saccharide onto a cation-exchange resin.

The K is defined as the ratio of the solute concentration in the resin phase $\overline{C}_S$ to that in the external solution phase C_S, and can be expressed by Eq. (1) based on the equality of the chemical potentials of the solute in the external solution and the resin phases [3,4].

$$K = \frac{\overline{C}_S}{C_S} = \gamma_0 \exp\left(-\frac{\Pi \overline{v}_S}{RT}\right) \tag{1}$$

where R is the gas constant, T is the temperature, $\overline{v}_S$ is the partial molar volume of the solute, and Π is the swelling pressure of the resin. γ_0 is a parameter reflecting both the ratio of the activity coefficient of the solute in the external solution phase to that in the resin phase, and the steric effect of the network of the resin frame on the distribution.

The binding constant B for complex formation of the solute and the counter-ion is defined by Eq. (2) according to the law of mass action, assuming that the activity coefficient of each species is unity and that a 1:1 complex is formed.

$$B = \frac{\overline{C}_{SM}}{\overline{C}_S \overline{C}_M} \tag{2}$$

where $\overline{C}_i$ (i = S, M or SM) is the concentration of the solute S, the counter-ion M or their complex SM in the resin phase.

Electroneutrality must hold in the resin phase, and is given by Eq. (3) when the concentrations of hydrogen and hydroxyl ions in the resin phase are negligibly low:

$$\overline{C}_{\mathrm{M}} + \overline{C}_{\mathrm{SM}} = \overline{C}_{\mathrm{E}} / z_{\mathrm{M}} \tag{3}$$

where $\overline{C}_{\mathrm{E}}$ is the concentration of fixed ions, which are considered to be univalent, in the resin phase, and z is the valency of the counter-ion.

The K_{app} is the ratio of the total concentrations of the free solute and the solute bound to the counter-ion in the resin phase ($\overline{C}_{\mathrm{S}} + \overline{C}_{\mathrm{SM}}$) to that of the solute in the external solution phase C_{S}, and is given by Eq. (4) from Eqs. (1) to (3).

$$\begin{aligned} K_{\mathrm{app}} &= \frac{\overline{C}_{\mathrm{S}} + \overline{C}_{\mathrm{SM}}}{C_{\mathrm{S}}} = K\left(1 + \frac{B\overline{C}_{\mathrm{E}} / z_{\mathrm{M}}}{1 + KBC_{\mathrm{S}}}\right) \\ &= \gamma_0 \exp\left(-\frac{\Pi \bar{v}_{\mathrm{S}}}{RT}\right)\left(1 + \frac{B\overline{C}_{\mathrm{E}} / z_{\mathrm{M}}}{1 + \gamma_0 \exp(-\Pi \bar{v}_{\mathrm{S}} / RT) BC_{\mathrm{S}}}\right) \end{aligned} \tag{4}$$

When C_{S} is sufficiently low, Eq. (4) can be approximated as follows:

$$K_{\mathrm{app}} = K(1 + B\overline{C}_{\mathrm{E}} / z_{\mathrm{M}}) = \gamma_0 \exp(-\Pi \bar{v}_{\mathrm{S}} / RT)(1 + B\overline{C}_{\mathrm{E}} / z_{\mathrm{M}}) \tag{5}$$

3. MATERIALS AND METHODS

3.1. Materials

A cation-exchange resin with sulfonate groups and a divinylbenzene content of 4% (Dowex 50W × 4) was converted into Ca^{2+} form accoding to standard procedures. D-Fructose, D-mannose and D-glucose were used as solutes. Dextran T-70, which possessed a molecular mass of about 7×10^5 and could not penetrate into the resin, was used for estimation of the bed voidage ε_b.

3.2. Apparent distribution coefficient

The resin particles were packed into a cylindrical glass column of 1.6 cm i.d. The bed height was about 35 cm, and was precisely measured in each experiment. A solute solution of a known concentration C_{S0} was continuously fed at *ca.* 1.0 mL/min, which was also precisely measured in each experiment, with a peristaltic pump. The effluent was fractionated at appropriate intervals Δt_{fra} (usually 2 or 3 min). The solute concentrations in the fractions were determined by a flow injection analysis with a fractometer to obtain the breakthrough curve. The amount adsorbed onto the resin q_S, which had the units of mass per unit volume of the resin phase, was evaluated from the mass balance equation (6) for the solute.

$$(1-\varepsilon_b)V_{bed}q_S = QC_{S0}t_E - \varepsilon_b V_{bed}C_{S0} - Q\Delta t_{fra}\sum_{i=1}^{N}C_{Si} \tag{6}$$

where C_{Si} is the solute concentration in the ith fraction, Q is the volumetric flow rate, V_{bed} is the bed volume, t_E is the time at the end point, and N is the number of fractions.

The apparent distribution coefficient K_{app} was calculated by Eq. (7).

$$K_{app} = q_S / C_{S0} \tag{7}$$

The solute concentration C_{S0} was in a range of 0.5% (w/v) to 40%.

After the bed was filled with the solute, a mixture of Dextran T-70 (0.5% (w/v)) and the solute, the concentration of which was C_{S0}, was continuously fed and the effluent was fractionated every 1 min. The concentration of Dextran T-70 in each fraction was determined by HPLC to obtain its breakthrough curve. The bed voidage was evaluated by analyzing the curve according to the similar equation to Eq. (6).

4. RESULTS AND DISCUSSION

4.1. Breakthrough curves

Figure 2 shows the breakthrough curves for glucose, mannose and fructose at two different concentrations. The figure also shows an example of breakthrough

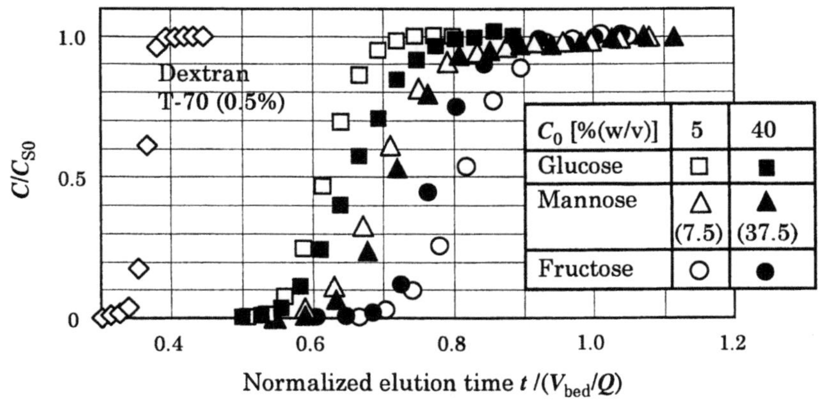

Figure 2. Breakthrough curves of glucose, mannose, fructose and Dextran T-70 in the bed packed with Dowex 50W×4 in Ca^{2+} form.

curves for Dextran T-70. Glucose eluted out earlier at C_{S0} = 5 % than at C_{S0} = 40%, indicating that its adsorbality onto the resin was higher at the higher C_{S0}. In contrast, the mean residence time of fructose was shorter at C_{S0} = 40% than at C_{S0} = 5%. Mannose, which eluted between glucose and fructose, exhibited no significant dependency in elution on its feed concentration.

Figure 3 shows the dependencies of the K_{app} values of glucose, mannose and fructose and the relative bed height Z/Z_0 on their concentrations. Although the bed voidage is not shown, it was almost constant (*ca.* 0.36) irrespective of the solute and its concentration. Therefore, the reduction in Z/Z_0 indicated the shrinkage of the resin itself.

Since the bed height was the same for all the solutes at $C_S \rightarrow 0$ and the molar volumes of the solutes v_S were common (0.114 L/mol) [2] (v_S was assumed to be used instead of $\bar{v}_S$), the difference in K_{app} at $C_S \rightarrow 0$ among the solutes reflected the difference in the B, as understood from Eq. (5). The B values of fructose, mannose, and glucose would be in this order, that is, fructose most strongly formed a complex with Ca^{2+} among the solutes.

When fructose was fed, no significant bed shrinkage was observed at any concentrations. This meant that the swelling pressure of the resin remained almost constant, and the intrinsic distribution coefficient K was constant over the concentration range tested. However, the K_{app} decreased as the C_{S0} increased. This can be explained from Eq. (4) as follows: the term KBC_S in the denominator of Eq. (4) became significant because of the large B value although the K value remained

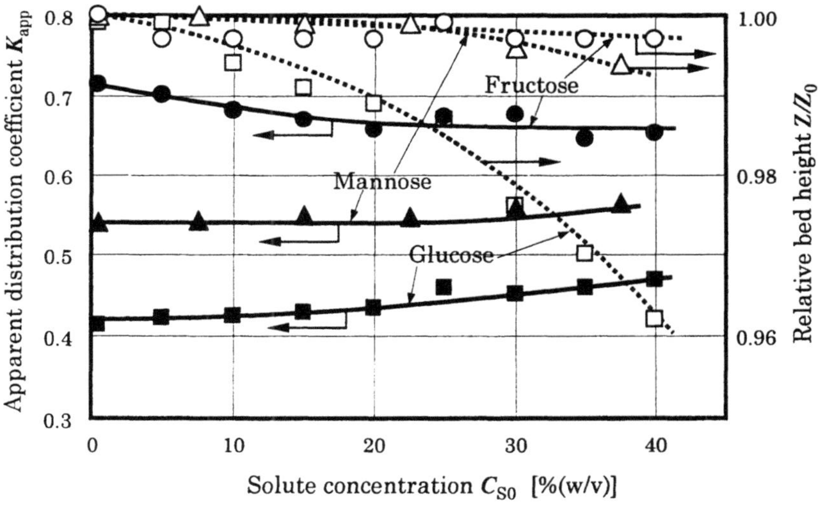

Figure 3. Dependencies of the K_{app} and the relative bed height Z/Z_0 of glucose, mannose and fructose on their concentrations.

constant.

On the other hand, the bed largely shrank, and the swelling pressure of the resin Π decreased as the glucose concentration became high. The decrease in Π resulted in the increase in the K as shown in Eq. (1). As mentioned above, the B value of glucose was the smallest among the solutes tested, and then the term KBC_S in the denominator of Eq. (4) would be insignificant over the concentration range. Therefore, the increase in the K_{app} with increasing C_{S0} could be ascribed to the increase in the K value.

The K_{app} of mannose scarcely depended on its concentration although the bed slightly shrank as the concentration increased. Mannose possessed the moderate B value. Therefore, the reason why the K_{app} remained almost constant could be explained by the setoff of the increase in the K value and the decrease in the term in parenthesis of Eq. (4) as the C_{S0} increased.

5. CONCLUSIONS

The K_{app} of glucose, which bound to Ca^{2+} most weakly among the hexoses tested, increased with increasing its concentration due to the reduction in the swelling pressure of the resin. On the other hand, the K_{app} of fructose, which bound to Ca^{2+} most strongly, decreased and the bed height did not change as its concentration increased. The K_{app} of mannose scarcely depended on its concentration because of the setoff of the reduction of the swelling pressure and the complex formation with the counter-ion.

Acknowledgement

We thank Mr. S. Narita and Ms. C. Furuta for their technical assistance.

REFERENCES

1. R. W. Goulding, J. Chromatogr., 103 (1975) 229.
2. S. Adachi and R. Matsuno, Biosci. Biotech. Biochem., 61 (1997) 1296.
3. S. Adachi, T. Watanabe and M. Kohashi, Agric. Biol. Chem., 53 (1989) 3203.
4. S. Adachi, T. Mizuno and R. Matsuno, J. Chromatogr. A, 708 (1995) 177.

Bioseparation Engineering
I. Endo, T. Nagamune, S. Katoh and T. Yonemoto (Editors)

Separation behavior of proteins near the isoelectric points in electrostatic interaction (ion exchange) chromatography

Takashi Ishihara and Shuichi Yamamoto*
Department of Chemical Engineering, Yamaguchi University
Tokiwadai, Ube 755-8611, Japan
*E.mail shu-yama@po.cc.yamaguchi-u.ac.jp fax +81-836-35-9933

Electrostatic interaction chromatography (EIC) commonly called ion exchange chromatography (IEC) is a very efficient and versatile method for separating and purifying proteins. Among various elution methods linear gradient elution (LGE) is most efficient for purifying a target protein from similar contaminant proteins. It is also useful for obtaining important information needed for understanding separation mechanisms rapidly. In this paper, our proposed method using LGE-IEC was described briefly, and applied to separation of β-lactoglobulin A and B (LgA, LgB) as model proteins. Resolution and retention of proteins in IEC near the isoelectric points (pI) were carefully investigated. The number of binding sites-pH relationships determined from LGE-IEC experiments were analyzed in order to understand good resolution (separation) near the pI.

1. INTRODUCTION

Electrostatic interaction chromatography (EIC) commonly called ion exchange chromatography (IEC) is a very efficient and versatile method for separating and purifying proteins[1-6]. A very simple net charge concept shown in Fig.1 is usually employed to explain the principle of IEC separation[5-6]. Although adsorption onto IEC supports can be roughly estimated on the basis of the net charge concept, it is not adequate for describing retention and separation of proteins in very high resolution IEC where both molecular recognition and transport phenomena play important roles[7-10].

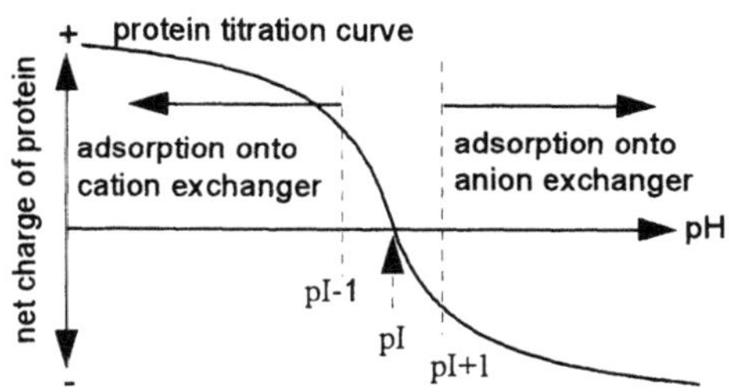

Fig.1 Relationship between pH and protein net charge charge

In this paper a fast and simple method for obtaining important information on molecular recognition, peak retention (and the number of binding sites) and peak resolution of proteins from linear gradient elution experiments is described. Linear gradient elution (LGE) is an elution method where the salt concentration is increased linearly at a fixed mobile phase pH to elute proteins initially loaded onto the column. The proposed method was verified with β-lactoglobulin A and B (LgA, LgB) as model proteins. The gradient elution experimental data were obtained over a wide range of mobile phase pH with various types of IEC media. The number of binding sites-pH relationships determined from LGE-IEC experiments were analyzed in order to understand good resolution (separation) near the pI. Effects of sample loading on retention and resolution near the pI were also investigated.

2. EXPERIMENTAL

2.1 Ion exchange chromatography media

Anion exchange media: Q-Sepharose HP, ANX-Sepharose HP (nominal particle diameter d_p=34μm), DEAE-Sepharose FF (d_p = 90 μm), Resource Q (d_p = 15μm), DEAE-Toyopearl 650S (d_p = 40 μm), TSK gel DEAE-NPR (d_p = 2.5 μm). Cation exchange media: SP-Sepharose HP, CM-Sepharose HP (d_p= 34μm), Resource S (d_p = 15μm), CM-Toyopearl 650S (d_p = 40 μm), TSK gel SP-NPR (d_p = 2.5 μm). Sepharose (agarose-based media) and Resource (stylene-divinylbenzene-based meida) are products of Amersham Pharmacia Biotech (Uppsala, Sweden). TSK-gel NPR (non-porous polymer particle) and Toyopearl (hydrophilic vynil polymer) media were supplied from Tosoh (Tokyo, Japan).

2.2 Materials

Bovine milk β-lactoglobulin(Lg) was obtained from Sigma (product no.L0130, St. Louis, MO), which contains both β-lactoglobulin Aand B. Other reagents were of analytical grade.

2.3 Chromatography experiment

Most experiments were performed on fully automated liquid chromatography systems, BIOCAD (Perseptive Biosystems, Boston, USA) and Prosys workstation (Beckman, Fullerton, USA)]. Sepharose gels were packed into a glass column (0.8cm ID and bed height Z= 15 cm) according to the recommended packing procedure [5]. Packed columns were checked by measuring HETP at non-binding conditions (pH 6 + 0.5M NaCl) using vitamin B12 as a tracer. Resource (0.64cm ID and Z=3cm) and NPR (0.46cm ID and Z=5cm) columns were supplied as packed columns. Buffer solutions were 10 mM acetate buffer (pH 4.0 - 6.0) or 10 mM Tris-HCl buffer (pH 7.0 - 9.5). The initial mobile phase solution (Solution A) was the buffer containing 0.03M NaCl and the final solution (Solution B) was the same buffer containing 0.5M NaCl. The gradient slopes g [M/mL=(mol/dm^3)/cm^3] were chosen so that baseline separation of LgA and LgB was attained. Therefore, the g values for DEAE-Sepharose FF were much lower than those for the other media (gradient volume = 4-32 column bed volume). The following experimental conditions were employed unless otherwise noted: the volumetric flow rate F= 0.5 mL/min for Sepharose FF, 1-8mL/min for Resource, 1 mL/min for Sepharose HP and 0.5 mL/min for NPR unless otherwise noted; the sample (Lg) concentration C_0 = 1mg/mL; the sample volume V_F= 0.5mL. The experiments were done at 298±1K.

3. RESULTS AND DISCUSSION

3.1 Short cut methods for obtaining the distribution coefficient by linear gradient elution experiments

Linear gradient elution(LGE)-IEC is an elution method, in which a linear increase of salt concentration is introduced to an IEC column inlet at a fixed pH after a sample protein is charged to the column. The charged protein is first strongly retained (adsorbed) to the column, and then gradually moves down the column after the introduction of the linear increase of salt concentration, I (See Fig.2). This is because the distribution coefficient K decreases sharply with increasing I. It is important to know $K(I)$ for predicting the peak retention time in LGE[4,8,9,12-13].

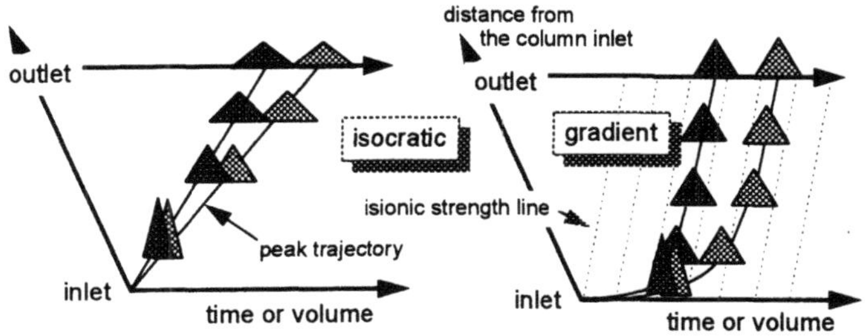

Fig.2 Schematical drawing of the movement of solute zones in the column in isocratic and gradient elution [4]

Also another important information can be extracted from $K(I)$. If we assume that the law of mass action (ion-exchange equilibrium) is valid [1-4,10-13,18-21] and K does not depend on the protein concentration C_P (low C_P or linear adsorption equilibrium), the following equation is derived.

$$K = K_e \Lambda^B I^{-B} \qquad (1)$$

where K_e is the equilibrium association constant, Λ is the (effective) total ion-exchange capacity, and B is the number of sites (charges) involved in protein adsorption, which is basically the same as the "Z" number in the literature [1-3, 10, 20]. Basically $K(I)$ can be obtained either by isocratic elution or by a batch experiment. However, there are several difficulties in obtaining $K(I)$ by these methods. First, there is no *a priori* information on $K(I)$ so that it is not easy to choose I arbitrarily. For example, If I is too low, the corresponding K value may be so large (say K>>10) that the peak can not be detected. $K(I)$ should be determined in the range of K =1 to 10. Secondly, even in the isocratic elution method it is not easy to detect a target protein peak if contaminants are included. In many cases the resolution of the target protein in isocratic elution is not good enough. On the contrary, in LGE-IEC most proteins are eluted in the range of I=0.03 to 0.5 M(=mol/dm^3). The resolution can be easily improved by decreasing the flow-velocity as well as by decreasing the gradient slope [4,17-18].

Our method for obtaining $K(I)$ or predicting the peak retention volume is explained briefly[4,12-13,17-18]. LGE-IEC experiments are performed at different gradient slopes g at a fixed pH. g is given by

$$g = (I_f - I_0)/V_G \quad \text{[M/mL]}$$

where I_f=final salt concentration, I_0=initial salt concentration and V_G=gradient volume. The salt concentration at the peak position I_R (see Fig.3) is determined as a function of the normalized gradient slope GH which is defined as

$$GH = gV_s = g(V_t - V_0) \qquad \text{[M]}$$

where V_t=total bed volume and V_o=void volume. The GH - I_R curve thus constructed does not depend on the flow-velocity, the column dimension, the sample loading (if it is not overloading conditions), or the initial salt concentration I_0 [4,12,17-18] as shown in Fig.4. It is recommended that GH values are in the range of 0.001 to 0.05. Usually the experimental GH-I_R data can be expressed by the following equation [4,12-13,18] (See.Fig.4).

$$GH = I_R^{(B+1)}/[A(B+1)] \tag{2}$$

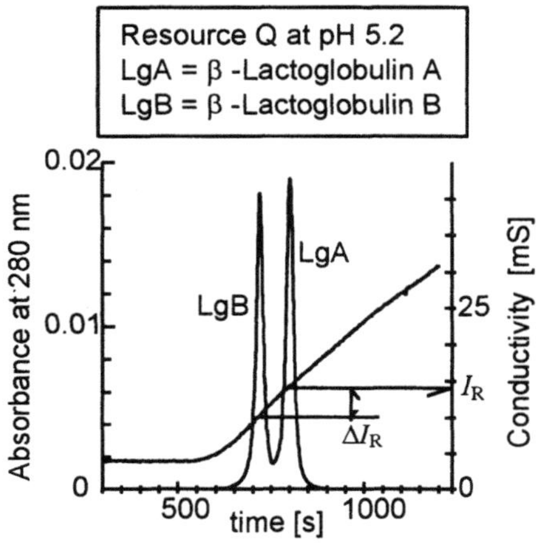

Fig. 3 Typical elution curves in linear gradient elution.

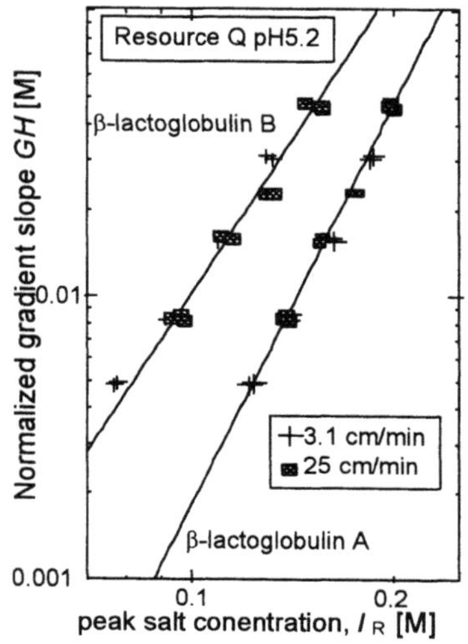

Fig.4 GH-I_R curves for β-lactoglobulins

From the law of mass action (ion-exchange equilibrium) [1-4,10-13,18-21], the following relationship can be derived.

$$A = K_e \Lambda^B \tag{3}$$

If we are only interested in predicting the peak retention in LGE-IEC, Eq.(1) can be used with A and B as experimental values [4,12,18]. However, if we construct the GH-I_R curves as a function of pH and determine the B - pH relationships, quite important information can be obtained on the retention (or molecular recognition) and the resolution of proteins as a function of pH. This is especially useful near the protein isoelectric point pI as many proteins behave very differently near the pI.

Application of the proposed method to separation of β-lactoglobulin A and B forms

As shown in Fig.3(typical elution curves, chromatograms) β-lactoglobulin A (LgA) and β-lactoglobulin (LgB) were separated on anion exchange chromatography (AIEC) columns at pH 5.2 although the degree of resolution varied from media to media (and depended on the operating conditions). The resolution became poor when the pH was increased from 5.2. The two proteins were not separated on any cation exchange chromatography (CIEC) columns used in this study at pH 4-5.6 even on the most efficient column (non-porous HPLC,

SP-NPR) at shallow gradient slopes.

The GH-I_R curves on a log-log scale shifted to larger I_R values and became steeper with increasing pH in the AIEC columns [4,12,13]. This implies that the number of adsorption sites B decreases when the pH approaches the pI, which is understandable in terms of the protein titration curve [1-6]. The slope increased with decreasing pH in the CIEC columns although the separation (resolution) of LgA and LgB was not observed under the conditions employed here as stated previously.

The B values determined from the GH-I_R curve as a function of pH are shown in Fig.5. When the pH for AIEC was increased, the B values increased and the difference between the B values of the two proteins decreased. Similarly, the B values in CIEC increases with decreasing pH from the pI. Even near the pI (pH 5.2) LgA and LgB were retained on both AIEC and CIEC columns. This can not be explained on the basis of a simple protein net charge concept shown in Fig.1. Although the peak salt concentration I_R value at a certain GH is different from media to media, the number of adsorption sites is not much different as shown in Fig.3 for DEAE-NPR and DEAE Toyopearl 650. The B values for LgB minus than those for LgA were *ca.* 1.0.

The resolution (separation) R_s values were highest around pH 5.2 (near the pI) with the AIEC columns. The R_s decreased very sharply with increasing pH for AIEC from pH 5.2 to 7.0. We have already proposed a dimensionless parameter $Y=[(ZD_aI_a)/(GH\ u\ d_p^2)]$ and shown that the resolution R_s can be tuned on the basis of this parameter[4,17]. Further, when R_s values at pH 5.2 for various AIEC columns are corrected as $R_s'[= R_s\ (\Delta I_{R,\mathrm{Resource\ Q}}/\Delta I_R)]$, R_s' were well correlated with Y (ΔI_R =($I_{R,LgA}$ - $I_{R,LgB}$) =peak salt concentration difference, $\Delta I_{R,\mathrm{Resource\ Q}}$ =ΔI_R for Resource Q column) [13].

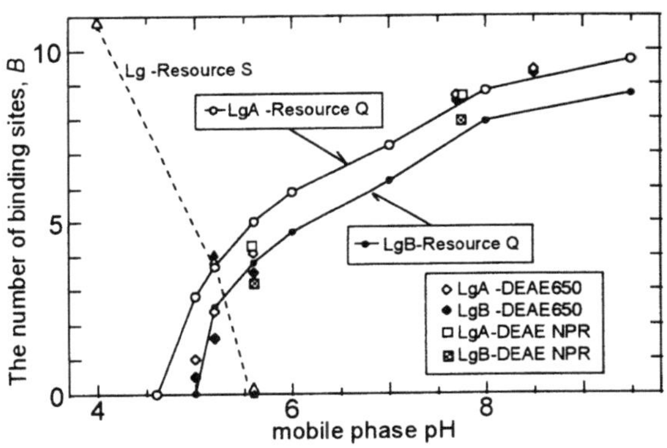

Fig.5 The number of binding sites as a function of mobile phase pH [13]

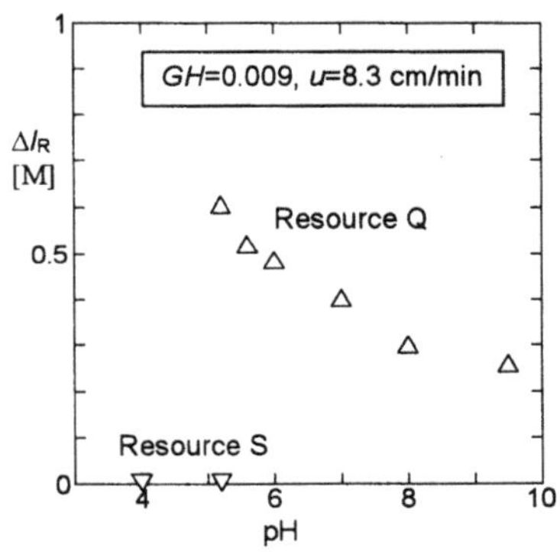

Fig.6 ΔI_R as a function of mobile phase pH

Fig.6 shows ΔI_R values (See Fig.3) as a function of mobile phase pH. The ΔI_R value did not depend very much on GH but decreased when pH was increased from 5.2 to 9.5. This decrease in ΔI_R with increasing pH is responsible for a decrease in R_s.

One of the concerns on IEC of proteins near the pI is that the solubility of proteins becomes low (precipitation may occur) and the resolution may be quite sensitive to sample loading. We measured the absorbance at 600nm (turbidity) of Lg, bovine serum albumin (BSA) and ovalbumin(OVA) for pH 4.8-6.0. BSA and OVA showed increases in turbidity when the concentration is in the range of 1 to 10 mg/mL. Lg was quite stable and did not show a remarkable increase in the turbidity even when the concentration is higher than 10 mg/mL at pH 5.2.

The effects of sample loading on the resolution R_s and the peak salt concentration I_R were examined. Both R_s and I_R values were constant up to 1 mg-protein/mL-bed. The R_s values at pH 5.2 were higher than those at pH 6.0 even when the sample loading is 20 mg/mL. The dependence of I_R on the sample loading at pH 5.2 was similar to or slightly weaker than that at pH 6.0.

Acknowledgment

This work was supported by a Grant-in Aid for scientific research on Priority Areas(No.296) (Grant No.11132255) and (C2, No.10650746) from the Ministry of Education, Science, Sports and Culture, Japan

REFERENCES

1. F.Regnier, Methods in Enzymol., 104,170(1984)
2. E.Karlsson, L.Ryden and J.Brewer, Ion-exchange chromatography, in Protein purification 2nd ed., p.14, ed. by J-C.Janson and L.Ryden, Wiley-VCH, 1998
3. G.Sofer and L.Hagel, Handbook of Process Chromatography, Academic Press, 1997
4. S.Yamamoto, K.Nakanishi and R.Matsuno, Ion-exchange chromatography of proteins, Marcel Dekker, New York, 1988
5. Ion-exchange chromatography Principles and Methods 3rd ed, Pharmacia Biotech, 1991
6. R.K.Scopes, Protein purification 2nd ed., Springer-Verlag, New York, 1987
7. M.Ladisch, Bioseparations, Kirk-Othmer Encyclopedia of Chemical Technology, 4th ed., Wiley, NY,Supplement, 89-122,1998
8. D.LeVan, G.Carta and C.M.Yon, Sec16 Adsorption and ion exchange,in Perry's Chemical Engineering Handbook,1997
9. G.Guiochon, S.G.Shirazi and A.M.Katti, Fundamentals of preparative and nonlinear chromatography, Academic Press, Boston, 1994
10. W.Kopaciewicz, M.A. Rounds, J. Fausnaugh and F.E. Regnier, J.Chromatogr., 266,3 (1983).
11. L.A Haff, L.G. Fagerstam and A.R. Barry, J. Chromatogr., 266, 409(1983).
12. S.Yamamoto, M. Nomura and Y.Sano, AIChE J, 33, 1426(1987) .
13. S.Yamamoto and T.Ishihara, J.Chromatogr. A, 852, 31(1999)
14. Piez, E.W.Davie, J.E.Folk and J.A.Gladner, J.Biol.Chem., 235, 2912(1961).
15. P.G.Righetti and T.Caravaggio, J.Chromatogr., 127,1 (1976) .
16. P.G.Righetti, G.Tudor and K. EK, J.Chromatogr., 220,115(1981).
17. S.Yamamoto, M. Nomura and Y.Sano, J.Chromatogr., 409, 101(1987).
18. S.Yamamoto, Biotechnol. Bioeng., 48, 444(1995).
19. N.K.Boardman and S.M. Partridge, Biochem. J., 59, 543(1955).
20. C.M.Roth, K.K.Unger, A.M.Lenhoff, J.Chromatogr. A, 726,45(1996).
21. S.R.Gallant, S.Vunnum and S.M.Cramer, J.Chromatogr. A, 725,295(1996)

Chapter 2

Refolding Processes for Protein

Bioseparation Engineering
I. Endo, T. Nagamune, S. Katoh and T. Yonemoto (Editors)

Large-Scale Refolding of Therapeutic Proteins

Jun Honda, Hidetoshi Andou, Teruhisa Mannen and Shunjiro Sugimoto

Bio-pharmaceuticals Development Center, Hoechst Marion Roussel Ltd.
3-2 Minamidai 1-chome, Kawagoe, Saitama 350-1165, Japan
E-mail: jun.honda@hmr.co.jp

Recombinant human growth differentiation factor 5 (rhGDF5) was refolded at remarkably high concentration of 2.4 mg/ml with an yield of 63%. After purification, the final yield at lab-scale was 20% with a purity of greater than 99%. The yield was twice that of conventional process having 3 chromatography steps and the purity was equivalent. The result of the first pilot-scale trial has shown a refolding yield of 51% and the final yield of 11%. This final yield is still 40% better than that of conventional process. Further optimization at pilot-scale is expected to bring these figures up to or above those of lab-scale, and will contribute significantly in reducing the production cost of rhGDF5.

Key words: Refolding, GDF5, Purification, Recombinant Protein, Large-Scale.

1. Introduction

Protein refolding is an important step in the downstream process of therapeutic protein production when the starting material is an insoluble inclusion body produced by over-expression in *E. coli*. This is also a step that is least understood in mechanism. Empirical methodologies have been accumulated over the years, and some standard protocols have been established (1). However there are still problems that we encounter when it comes to applying this step in large scale, such as low yield per volume of reaction mixture, low yield of refolded protein, high cost of reagents used, etc.

Hoechst Marion Roussel has a number of therapeutic proteins that are on the market or under development. Among them, recombinant human insulin had a success in the development of refolding step at large scale (2). In this paper, we describe an improved production process developed for industrial scale that is extremely efficient for recombinant human growth differentiation factor 5 (rhGDF5).

2. rhGDF Production: Present Status

rhGDF5 is one of bone morphogenetic proteins (3), belonging to transforming growth factor β (TGF-β) superfamily. It has commercial value as a therapeutic protein because of its ability to induce cartilage and bone formation (4), and angiogenesis (5) in adult animals. Recombinant form of human growth differentiation factor 5 (rhGDF5) (6) is a homodimer of 119 amino acid residues per monomer with an approximate molecular weight of 26,000 and has no cofactors. It is produced by over-expression in *E. coli* in a form of insoluble inclusion body. Presently, rhGDF5 is purified by solubilization of inclusion body by 8M urea and dithiothreitol (DTT); subjected to 2 chromatographic steps (ion exchange and gel filtration) under the presence of 6M urea; refolded by dilution in the presence of oxidized glutathione and zwitterionic detergent CHAPS; and then purified with a final reverse-phase chromatography step. This established process (6) however, is relatively costly and this is due mainly to the property of rhGDF5 being very insoluble and interactive with chromatographic media. The problems associated with this process is shown in Table 1.

Improvement of this process has been sought, and the concept of the new process is to bring the refolding step right at the beginning, since it is more reasonable to purify the refolded product (dimer) rather than its monomeric precursor. However, this would mean that the refolding step will be the critical step in the whole downstream process of rhGDF5 production and also that the process development will be challenging due to the above-mentioned property of rhGDF5.

3. Process Optimization

With the proteins that involve disulfide bond formation during refolding, addition of

Table 1: Problems associated with rhGDF5 purification process (per 700 L fermentation batch)

- Amount of urea used (2800 kg/batch)
- Amount of redox reagents used (18 kg/batch)
- Large refolding vessel
- Costly gel filtration media
- Low throughput (4 months/7 batches)

oxidizing reagents is essential to counteract the effect of reducing reagent which is added during the solubilization step to "unscramble" the spontaneously formed disulfide bonds. In many cases, the reagents of choice are mixtures of reduced and oxidized glutathione (7). They are however, too expensive for industrial use. So instead of dithiothreitol/glutathione combination used in the conventional method, the effect of introducing cysteine was tested for the new process. This proved to be quite effective for the refolding of rhGDF5. After addition and dilution of cysteine, it seems that the reducing potential of free cysteines gradually decreases with time, providing an oxidative environment favorable for rhGDF5 refolding. Therefore, an addition of oxidizing reagent became unnecessary.

CHAPS and NaCl were indispensable for formation of dimer, and urea and arginine were indispensable for suppressing aggregation to increase solubility of rhGDF5. The optimized refolding condition is shown in Table 2.

In order to subject mature rhGDF5 to chromatographic steps, CHAPS has to be removed because its surface active property disturbs rhGDF5 from binding to hydrophobic reverse-phase media. In the conventional process, the recovery is performed by simple dilution and simultaneous precipitation at isoelectric point (pH 7.4), and then centrifugation to recover the precipitant. For large-scale production, diafiltration using ultrafiltration membrane is more suitable. This way, CHAPS are removed and the liquid volume is decreased simultaneously, thus decreasing the size of reaction vessel dramatically. Isoelectric precipitation is performed and the precipitate is recovered in the same manner as the conventional method. The recovered precipitate is then solubilized in phosphoric acid solution for further processing on reverse-phase chromatography, again in the same manner as the conventional method. The comparison of the two methods are shown in Figure 1.

Table 2: Refolding condition of the new method

2.4 mg/mL solubilized inclusion body
0.5 M Arginine-NaOH (pH 8.9)
0.5 M NaCl
20 mM CHAPS
4.8 mM Cysteine-HCl
0.75 mM EDTA
2.4 M Urea

Fig. 1: Comparison of the two methods

Conventional	New
Solubilization	Solubilization
▼	▼
Ion exchange	Refolding
▼	▼
Gel filtration	Diafiltration
▼	▼
Refolding	Isoelectric precipitation
▼	▼
Isoelectric precipitation	Reverse-phase
▼	
Reverse-phase	

Table 3: Comparison of yields between different processes

Step	Conventional* Pilot [%]	New Lab [%]	New Pilot [%]
Solubilization	100	100	100
Refolding	36	63	51
Preparative HPLC	64	43	26
Overall final yield	8	20	11

*Average of 3 batches

4. Process Scale-up

The established lab-scale process (8) was scaled-up by a factor of 70, based on the volume of refolding solution (from 2 L to 140 L). The result is shown on Table 3. Since the two processes are different, only step yields that can be directly compared and the final overall yields are shown. In the refolding step, the lab-scale trial of the new method showed step yield of 63% while its first pilot-scale trial showed 51%. This is 1.4 times better than the conventional average of 36%. On the other hand, step yield at reverse-phase HPLC column (bed volume: 13 L) was 26%, which is 60% of the lab-scale result, and 40% of the conventional process. This is due to narrow fraction pool due to increased impurity components in the new process. As a result, the final yield of new pilot trial was 11%, which is about 1.4 times better than the conventional pilot.

As for product quality, it was evaluated using the following items: SDS-PAGE, IEF, analytical HPLC, *in vitro* bioassay, peptide mapping, amino acid composition and N-terminal and C-terminal assay. These items showed that the quality of products between those produced by conventional and new processes are equivalent. For impurity profiles, contents of endotoxin and cell-derived proteins (mock) were examined. It was shown that endotoxin content was equivalent between the two processes, but that of mock (measured using ELISA) showed a higher figure. Mock from the conventional process was 2.09×10^1 ng/mg-rhGDF5, while that from the new process was 5.04×10^3 ng/mg-rhGDF5. Improvement of this figure will be one of the targets in the optimization trials to come.

As a result of product process modification, problems listed in Table 1 can be alleviated in a following manner. The amount of urea used can be reduced from 2800 kg to 170 kg per fermentation batch. That of redox reagents can be reduced from 18 kg to 1 kg likewise. Throughput can be increased from performing 7 purification batches per fermentation batch taking about 4 months, to 1 purification batch plus 4 HPLC sub-batches taking about 2 weeks. The calculated production cost reflects this difference. Assuming that rhGDF5 is produced for 10 years after launch in the same facility and taking all items such as personnel costs, energy costs, depreciation costs, etc. into consideration, it was shown that the cost per mg of rhGDF5 product can be reduced by 42%. This will no doubt increase market value of this product dramatically.

5. Conclusion

We have shown that a new process developed for the production of rhGDF5 is effective in reducing the cost, increases throughput substantially, is easy to scale-up, and is more friendly to the environment (because much less raw material is used). Optimization trials hopefully will show positive results and contribute in efficient production of rhGDF5 in the future, and also become basis for the production process of other recombinant therapeutic proteins to come.

6. References

(1) R.Rudolf, and H. Lilie, FASEB J., 10, 49-56 (1996)
(2) R. Obermeier, *et al.*, German Patent Application no. 4405179.4 (1994)
(3) G. Hoetten, *et al.*, Biochem. Biophys. Res. Commun., 204, 646-652 (1994)
(4) G. Hoetten, *et al.*, Growth Factors, 13, 65-74 (1996)
(5) H. Yamashita, *et al.*, Exp. Cell Res., 235, 218-226 (1997)
(6) F. Makishima, *et al.*, Patent Application no. WO9633215 (1996),
(7) B. Fischer, *et al.*, Biotechnol. Bioeng., 41, 3-13 (1993)
(8) J. Honda, *et al.*, Patent Application no. WO9829559 (1998)

Bioseparation Engineering
I. Endo, T. Nagamune, S. Katoh and T. Yonemoto (Editors)

Novel method for continuous refolding of protein with high efficiency

S. Katoh and Y. Katoh

Graduate School of Science and Technology, Kobe University
Nada-ku Rokkodai, Kobe 657-8501, Japan

Denatured and fully reduced lysozyme was refolded in batch, fed-batch and continuous operations. In fed-batch refolding of lysozyme, denatured lysozyme was gradually added into refolding buffers containing urea in the concentration range from 1.0 to 2.0 mol/L. The recoveries of the lysozyme activity in fed-batch operation were higher than those in batch operation. In continuous method, the denatured lysozyme solution was gradually added from the outer surface of the membrane tube into a refolding buffer flowing continuously inside the tube under controlled mixing conditions. The refolding efficiencies of lysozyme in this continuous refolding were higher than those in the batch operation.

1. INTRODUCTION

Recombinant proteins over-expressed in *Escherichia coli* are often accumulated as insoluble particles called inclusion bodies. Since proteins in inclusion bodies are usually inactive, they must be solubilized by a denaturing agent such as 8 mol/L urea or 6 mol/L guanidine HCl and refolded to recover their native steric structure having biological activities. In refolding process a solubilized protein solution is added into a large volume of a refolding buffer in order to reduce the concentration of a denaturing agent and also to avoid aggregate formation of protein molecules in the course of renaturation. Thus, a large volume of a stirred tank is required, and the concentration of proteins after renaturation becomes low. Further, difficulties in uniform mixing in large-scale stirred tanks cause heterogeneity in refolding conditions and reduce the efficiency of refolding.

In our previous work [1], fully reduced hen egg-white lysozyme was gradually added into a refolding buffer in fed-batch manner, and the performance of this refolding operation was higher than that of batch operation. On the other hand, continuous refolding operations have some advantages, especially in large-scale processes, such as high throughput, homogeneity in quality of refolded protein and also flexibility to meet various requirements for efficient refolding. They, naturally, can make overall downstream processes continuous. In the present work, in order to realize the advantages of both fed-batch addition and continuous operation, the effects of refolding conditions in fed-batch operation were studied, and a continuous refolding method with addition of denatured protein solutions in fed-batch manner is developed and applied to refolding of lysozyme.

2. MATERIALS AND METHODS

2.1. Materials

Hen egg-white lysozyme (MW 14,300, 6 x crystallized, Seikagaku-Kogyo Co.) and *Micrococcus lysodeikticus* dried cells (Sigma Chemicals Co.) were used. Other reagents used were of analytical grade.

2.2. Protein denaturation

A denaturation buffer (0.1 mol/L Tris-HCl, 8 mol/L urea, 10 mmol/L dithiothreitol (DTT), 1mmol/L EDTA, pH 8.5) was incubated under nitrogen stream for 30 min. Lysozyme was added to the buffer at concentrations of 5 - 35 kg/m^3, and fully denatured under nitrogen stream at 38 °C for 2 hrs. Complete unfolding of lysozyme was confirmed by the CD spectrometer and the number of free SH residues measured by Ellman's method.

2.3. Measurement of enzyme activity

In the measurement of lysozyme activity *Micrococcus lysodeikticus* dried cells were suspended (0.2 kg-solid/m^3) in a sodium phosphate buffer (50 mmol/L, pH 6.2). The absorbance of this substrate solution was about 1.0 at 450 nm (optical length :1 cm). The enzyme reaction was initiated by adding 3×10^{-9} m^3 of a sample to 1×10^{-6} m^3 of the substrate solution, and the decrease in absorbance was recorded continuously with a spectrophotometer (Shimadzu UV-1600) at 35 °C. The activity of lysozyme was determined from the initial slope.

2.4. Refolding of lysozyme

Batch operation

The denatured lysozyme solution was diluted 10-fold under stirring with a refolding buffer (0.1 mol/L Tris-HCl, 3 mmol/L reduced glutathione (GSH), 5 mmol/L oxidized glutathione (GSSG), 1 mmol/L EDTA, pH 8.0) containing 1.5 mol/L urea at room temperature. The total volume of the renaturation mixture was 4×10^{-5} m^3. The concentration of lysozyme in the renaturation mixture was determined from the absorbance at 280 nm measured immediately after dilution. The renaturation mixture was stirred for 24 hrs at room temperature, and then the enzyme activity of the sample was measured. The refolding efficiency of lysozyme was defined as the activity of the sample relative to that of the control solution with the same lysozyme concentration.

Fed-batch operation

The denatured lysozyme solution (10 - 30 kg/m^3) was continuously supplied with a micro-feeder pump (Furue Science Co., JP-V-W) at a flow rate of from 3.3×10^{-8} m^3/min to 2.6×10^{-7} m^3/min into 3.6×10^{-5} m^3 of the refolding buffer gently stirred. The time required for addition of 4.0×10^{-6} m^3 of the denatured lysozyme solution was ranged from 15 min to 120 min. After supplying 4.0×10^{-6} m^3 of the denatured lysozyme solution (ten times dilution and the final concentrations ranged from 1.0 to 3.5 kg/m^3), the concentration of lysozyme in the renaturation mixture was determined from the absorbance at 280 nm. The renaturation mixture was stirred further for 24 hrs at room temperature, and then the enzyme activity of the sample was measured.

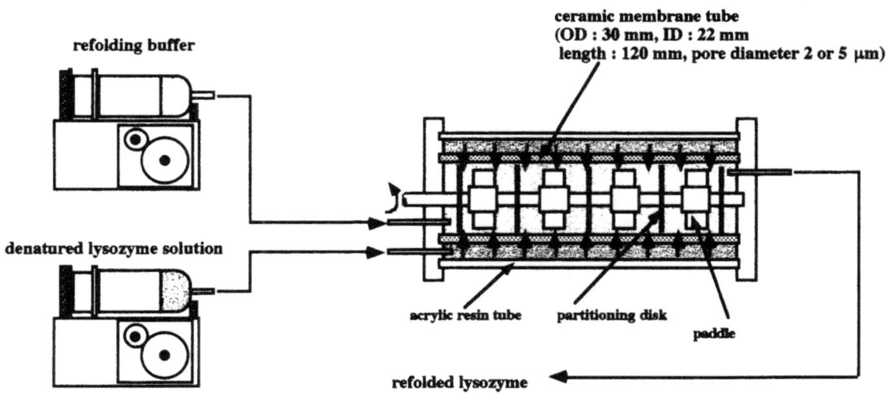

Figure 1 Continuous refolding apparatus

Continuous refolding with addition of denatured protein solution in fed-batch manner

A continuous refolding apparatus is schematically shown in **Figure 1**. The inner tube was made of ceramic membrane. The denatured lysozyme solution was supplied to the annular region by the micro-feeder pump at a flow rate of 1.35×10^{-7} m^3/min, and was forced to permeate to the inside of the tube through the membrane, because the annular space was dead-ended. The flux through the membrane was considered to be uniform throughout the surface of the membrane. The refolding buffer was supplied continuously to the inner tube by the micro-feeder pump at a flow rate of 1.22×10^{-6} m^3/min. The ratio of the flow rate of the denatured lysozyme solution to that of the refolding buffer was 1 : 9. The apparatus was equipped with four two-blades paddles and five partitioning disks (diameters : 20 mm) inside the tube, and the solution flowing in the tube was mixed under a controlled axial dispersion. The concentration of lysozyme in the refolding solution should gradually increase from the inlet to the exit because of the inflow of the denatured lysozyme solution through the membrane. The average residence time of the solution flowing in the tube was 40 min. The samples flowed out from the tube were collected at predetermined time intervals, and their lysozyme concentrations were determined from the absorbance at 280 nm. The samples were stirred further for 24 hrs at room temperature, and then their enzyme activities were measured.

2.5. Measurement of fluorescence intensity

Conformation changes of lysozyme during refolding process were monitored by the change in fluorescence intensity emitted from tryptophan residue (Em at 345 nm) of the sample with a spectrofluorophotometer (Shimadzu RF-540, Ex 280 nm, Em 250-500 nm).

2.6. Measurement of residence time distribution of solution in membrane tube

To determine the axial dispersion of solutions flowing inside the membrane tube, a step response was measured by application of a step input of blue dextran 2000 (Amersham Pharmacia Biotech.) solution and measuring continuously the absorbance of the exit stream

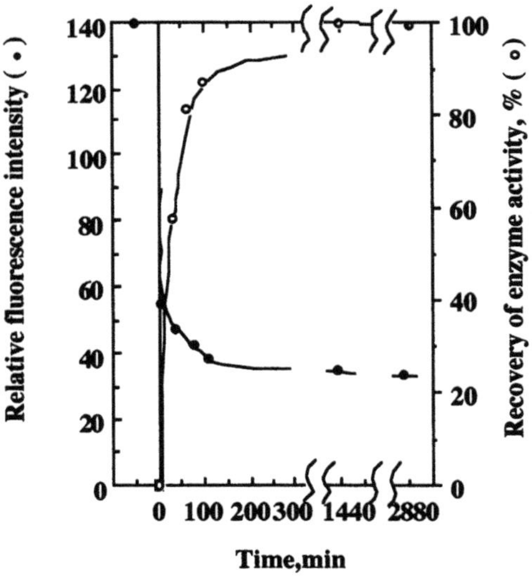

Figure 2 Recovery of enzyme activity and fluorescence intensity
lysozyme concn : 0.5 kg/m^3

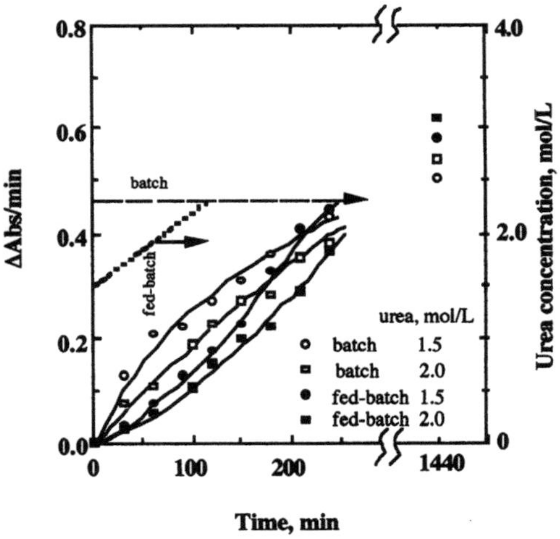

Figure 3 Time-courses of recovery of lysozyme activity
lysozyme concn : 2 kg/m^3

from the membrane tube at 620 nm. The step responses were analyzed by the tanks-in-series model [2], and the number of tanks corresponding to the axial dispersion of the solution in the tube was determined.

3. RESULTS AND DISCUSSION

3.1. Refolding of lysozyme in batch and fed-batch operations

In **Figure 2,** the recovery of lysozyme activity and the relative fluorescence intensity are plotted against time after mixing with the refolding buffers containing 1.5 mol/L urea in batch operation. The recovery and the fluorescence intensity became almost constant after one day, and thus the recoveries of the enzyme activity of lysozyme in the refolding buffer were compared after one day.

In **Figure 3**, the activity of lysozyme is plotted against time after mixing with the refolding buffer for batch operation and against time from the start of addition for fed-batch operation. In the initial stage, the rates of increase in the activity were lower in the cases of the higher urea concentrations in both batch and fed-batch operations and lower in the cases of fed-batch operation than batch operation. In fed-batch operation, lysozyme was gradually added into the refolding buffer for two hours, and hence the recovery of activity was naturally lower than that in batch operation. After one day, however, the recoveries of the enzyme activity in fed-batch operation were higher than those at the same lysozyme concentration in batch operation.

As shown with a broken line, the concentration of urea in the refolding buffer was a constant value of 2.3 mol/L in batch operation with the refolding buffer containing initially

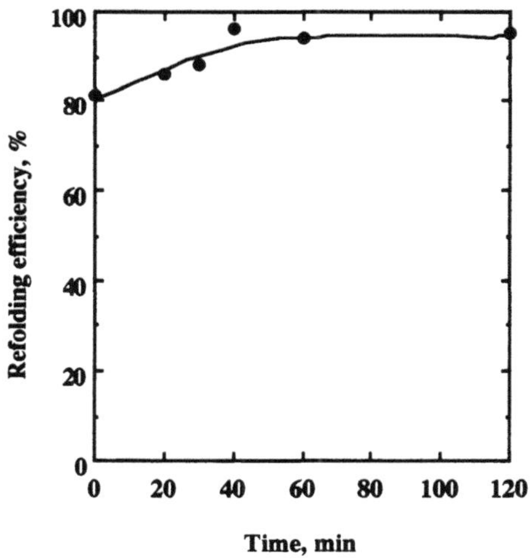

Figure 4 Effect of rate of addition
lysozyme concn : 1.5 kg/m^3

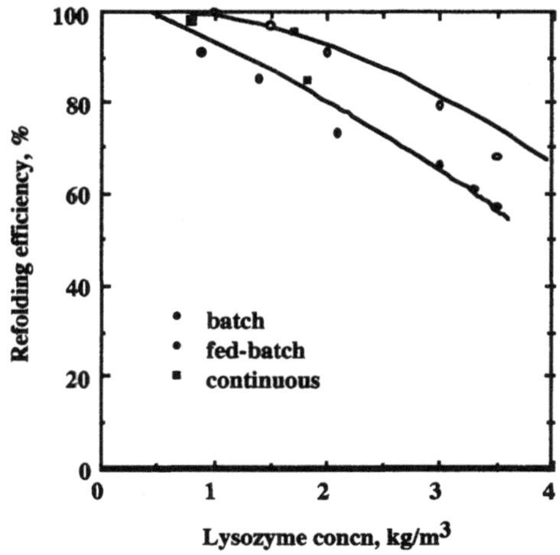

Figure 5 Comparison of refolding efficiencies in batch, fed-batch and continuous operations

1.5 mol/L urea, since denatured lysozyme in the denaturation buffer containing 8 mol/L urea was diluted ten times by the refolding buffer. On the other hand, as shown with a dotted line, the concentration increased linearly from 1.5 mol/L to 2.3 mol/L during 120 min for fed-batch operation. Thus, the effects of the concentration of urea in the refolding buffer should be different between batch and fed-batch operations. Higher refolding efficiencies were obtained with the refolding buffers containing 1.5 mol/L urea in fed-batch operation. Therefore, in this work the refolding buffer containing 1.5 mol/L urea was used throughout all experiments of batch dilution, fed-batch and continuous operations.

In **Figure 4** the refolding efficiency of lysozyme was plotted against the time needed for fed-batch addition of 4 x 10^{-6} m^3 of the denatured protein solution into 3.6 x 10^{-5} m^3 of the refolding buffer. The final concentration of lysozyme in the refolding buffer was about 1.5 kg/ m^3. The point on the y axis corresponds to the refolding efficiency with instantaneous mixing, i.e. with batch dilution operation. The refolding efficiency increased with decrease in the rate of addition, which means that lysozyme which is in the course of refolding in the fed-batch operation affects the refolding efficiency to some extent. The degree of increase in the efficiency, however, gradually decreased with increase in the time, ant therefore fed-batch addition of the denatured protein solution during 40 min was adopted for most cases.

Figure 5 compares the results of lysozyme refolding in the batch and fed-batch operations by use of the refolding buffer solutions containing 1.5 mol/L of urea. The recoveries of the lysozyme activity in fed-batch operation were higher than those in batch operation. By use of the refolding buffer containing 1.5 mol/L of urea, very high recoveries of the lysozyme activity were obtained in fed-batch operation. In the fed-batch operation, refolding of lysozyme started from lower concentrations, at which the recovery of the enzyme activity was 100 %, in case the concentration of urea in the refolding buffer was adjusted at a suitable value. With addition of the denatured lysozyme solution, lysozyme gradually accumulated in the refolding buffer solution. Since the refolding efficiency decreases with the concentration of lysozyme, the portion of lysozyme refolded at lower concentrations contributed to higher overall efficiency in the fed-batch operation. In case the refolding efficiency of protein decreases with the concentration of protein in the refolding solution, gradual addition of the denatured protein solution in the fed-batch operation results in refolding with high efficiency in the lower range of concentrations and thus, the overall efficiency will be improved.

3.2. Refolding efficiency of lysozyme in continuous refolding

In **Figure 5** the refolding efficiencies of lysozyme in continuous operations were also shown. In the continuous operation, the four paddles were rotated at 40 r.p.m. with 5 partitioning disks. The refolding efficiencies were almost same in fed-batch and continuous operations, while those of batch operation were lower.

Although the recovered activity of lysozyme was reached a plateau after 24 hrs, fed-batch addition of the denatured lysozyme solution during 15 - 120 min could improve the refolding efficiency. This means that renaturability of each denatured lysozyme molecule is assigned short after addition to the refolding buffer. Thus, the continuous refolding apparatus with the membrane tube was considered to act as an effective mixer of the denatured and refolding solutions in fed-batch manner and, therefore, might have an ability to show a high refolding efficiency and a high throughput with a small holding volume.

REFERENCES

1. S. Katoh, M. Terashima, H. Kishida and H. Yagi, J Chem. Eng. Japan, 30 (1997) 964.
2. O. Levenspiel, Chemical Reaction Engineering, pp. 290-295. John Wiley & Sons, New York, 1972.

Bioseparation Engineering
I. Endo, T. Nagamune, S. Katoh and T. Yonemoto (Editors)

Novel protein refolding by reversed micelles

M. Goto, T. Fujita, M. Sakono, and S. Furusaki
Department of Chemical Systems and Engineering,
Graduate School of Engineering,
Kyushu University, Fukuoka 812-8581, Japan

A novel protein refolding technique utilizing reversed micelles has been developed. Denatured RNase A solubilized in the reversed micellar solution was reactivated by addition of glutathione entrapped in another reversed micellar solution. This novel refolding method facilitates a high renaturation yield at high protein concentration compared to achievable results in dilution method. Novel refolding technique by reversed micelles makes it possible to activate proteins from insoluble aggregates by direct dissolution into the reversed micellar solutions.

1. INTRODUCTION

Proteins from eukaryotic genes cultivated in a bacterial system by recombinant DNA technology are often recovered in the form of inactive misfolded or aggregated proteins. In order to obtain biologically active proteins, these insoluble proteins must first be extracted from the inclusion bodies and solubilized using a denaturant. On the removal of the denaturant by dialysis or dilution with an appropriate buffer, the proteins are allowed to refold up to the native and active conformation. However, high refolding yield is not obtained at high protein concentration due to the formation of irreversible aggregates by the excess inactive protein molecules. A novel preparation method aimed at improving the efficiency of refolding at higher protein concentration was recently reported[1,2]. Reversed micelles provide nano-structural aqueous droplets constituting water pools in an organic solvent, so that proteins solubilized in the water pool can maintain their high-structure conformation and activity. These features are favorable for protein refolding, because protein molecules can be isolated from each other during the refolding process.

In the present study, we demonstrate that the reversed micellar technique is useful and attractive for protein refolding. Some key operational parameters were investigated regarding the efficiency of protein refolding, and the optimal conditions are discussed.

2. MATERIALS AND METHODS

2.1 Materials. Bovine pancreatic RNase A (EC 3.1.27.5) as a model protein and cytidine 2 :3 -cyclic monophosphate as a substrate for the activity test were obtained from Sigma Chemicals Co. (USA). AOT (Sodium bis-2-ethylhexyl sulfosuccinate), urea and ε-mercaptoethanol were purchased from Kishida Chemicals Co. (Japan). Reduced and oxidized glutathione (GSH, GSSG) were purchased from Wako Pure Chemical Industries, Ltd. (Japan).

2.2 Extraction of Denatured RNase A into Reversed Micelles. A reversed micellar solution was prepared by injecting a 0.1 M Tris-HCl buffer (pH= 7.6) into a 50-400 mM AOT-isooctane solution. The solid denatured RNase A was dissolved into the reversed micellar solution under ultrasonic irradiation for approximately 30 minutes. The concentration of RNase solubilized in the reversed micellar solution was measured from the absorbance at 278 nm with a UV-VIS spectrophotometer (U-best 570 JASCO). The average diameter of reversed micelles containing proteins was analyzed by a DLS-7000 system.

2.3 Renaturation of RNase A in Reversed Micelles. Renaturation was initiated by adding 1 ml of reversed micellar solution containing glutathione into the 2 ml reversed micellar solution containing solubilized RNase A. The specific activity of RNase A in the reversed micelles was measured against time progress[3)]. The renaturation yields of denatured RNase A was evaluated by comparing the enzymatic activity of denatured RNase A to that of native RNase A which was solubilized in the reversed micelles as a control experiment.

2.4 Recovery of Renatured RNase A from Reversed Micelles

One volume of reversed micellar solution containing the renatured RNase A was added into 5-40 volumes of cold acetone and stirred in a magnetic stirrer for 20 minutes. The reprecipitated RNase A was collected by centrifugation and washed several times with cold acetone to remove AOT molecules. Furthermore, the precipitates were dried under vacuum condition. The obtained precipitates were solubilized in a 0.025 M phosphate buffer (pH 7.5). The amounts of recovered RNase A were determined by measuring the absorbance in the buffer solution at 278 nm with the UV-VIS spectrophotometer.

3. RESULTS AND DISCUSSION

The novel protein refolding process utilizing reversed micelles is schematically illustrated in Figure 1. It consists of three steps: solubilization, renaturation and recovery steps.

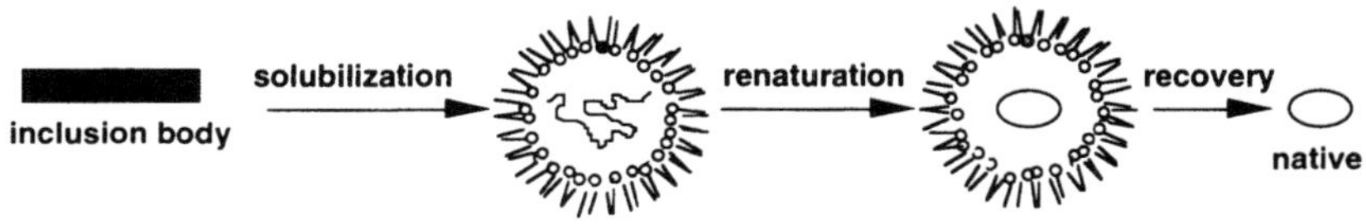

Figure 1. Schematic illustration of protein refolding by reversed micelles.

3.1 Solubilization of Denatured RNase A into Reversed Micellar Solution

In the previous study[1] protein refolding utilizing reversed micelles showed that solubilization of denatured RNase A into reversed micelles was difficult owing to a large amount of denaturant (6M GuHCl or 8M urea) in aqueous solution, which interrupted the mass transfer of proteins.

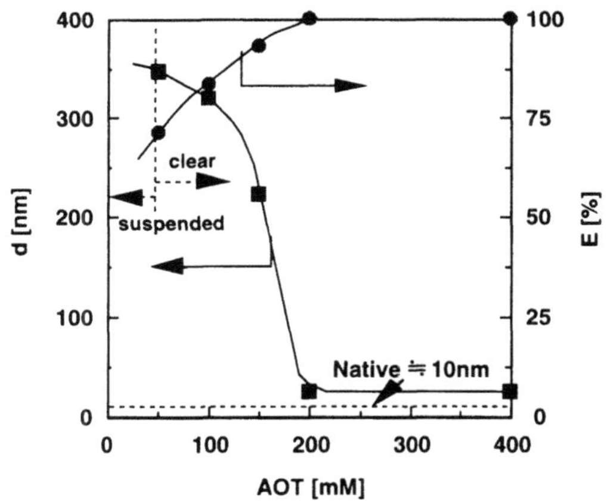

Figure 2. Effect of AOT concentration in isooctane on both solubilization and average diameter of reversed micelles after solubilization: 1 mg/ml RNase A, Wo (=[H2O]/[AOT]) =20, pH 8.7.

The denaturant reduces the electric interaction between surfactants and the surface of proteins, which is the main driving force in protein transfer by reversed micelles[4]. In order to improve the overall yield in the protein refolding process, we have introduced a solid-liquid extraction technique to enhance the solubilization of unfolded proteins directly into reversed micelles [3,4]. The solubilization of denatured proteins into the reversed micelles was completely performed in a few minutes. When the surfactant-protein interaction is stronger than the protein-protein interaction in the denatured proteins, the protein solubilization into the reversed micelles proceeds well. From the result of DLS (Dynamic Light Scattering) measurement, it was found that the average diameter of the stable reversed micelles containing denatured RNase A in the organic solvent was approximately 20 nm. On the other hand, under the same experimental condition the average diameter of reversed micelles that contain native RNase A was about 10 nm. This result means that the solubilization state of the denatured proteins is slightly different from that of native proteins. The denatured proteins might form a small cluster in reversed micelles.

3.2 Improvement of Solubilization Efficiency of Denatured Proteins by Urea Addition

To enhance the solubilization efficiency of denatured proteins, we improved the solubilization method. As a pretreatment, a small amount of concentrated urea solution (8M) was added to the solid denatured protein, then the turbid protein solution was dispersed into reversed micellar solution. The final urea concentration was adjusted to be 2 M on the basis of the total volume.

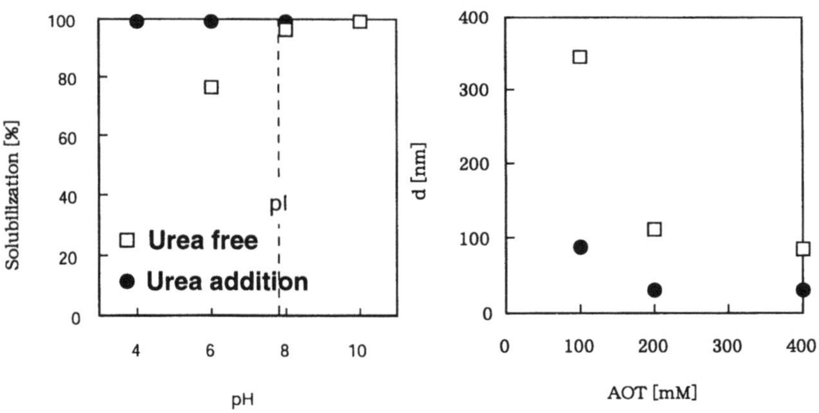

Figure 3. Effect of urea addition on the solubilization of denatured protein.

Figure 3 shows the effect of urea addition on the solubilization efficiency of denatured protein. It is surprised that the denatured protein was completely solubilized in reversed micelles regardless the pH conditions in the protein solution. The size of reversed micelles considerably reduced by the pretreatment.

According to circular dichroism (CD) spectra, it is obvious that the conformation of denatured RNase A entrapped in the reversed micelles remained in denatured states (data not shown) and the activity was almost zero at this stage. This implies that reoxidation of reduced thiols is essential to obtain native conformation and to recover the enzymatic activity. Hence, we used reduced and oxidized glutathione (GSH and GSSG) which is the most widely used compound as the thiol/disulfide reagents.

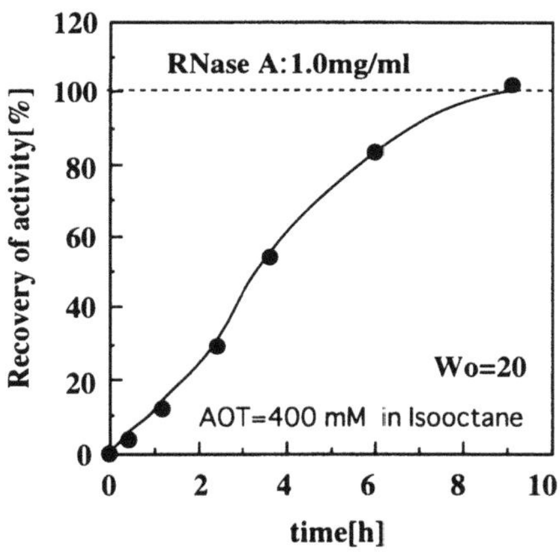

Figure 4. Time course of protein refolding in reversed micellar solutions.

Visible protein aggregates were not observed during refolding operations under the present experimental conditions. Renaturation of RNase A was conducted by the addition of another reversed micellar solution containing glutathione. Enzyme activity was recovered completely within 10 hours (Fig. 4). Thus, the efficiency of refolding is enhanced by using reversed micellar media at higher protein concentration. On the basis of this result, it is clear that the denatured protein molecules can be isolated from each other in the nano-structural molecular assemblies of reversed micelles and efficiently refolded without forming protein aggregations.

3.3 Recovery of Renatured RNase A from Reversed Micelles

Denatured RNase A has now been successfully reactivated in reversed micellar solution. The final step is to recover the active proteins from the reversed micellar solution. The recovery of proteins solubilized in reversed micelles has been performed by back extraction operation. In general, the reversed micellar solution is made contact with a fresh aqueous solution with a high ionic strength and a pH value that facilitates electrostatic repulsion between the encapsulated protein and the head group of surfactant molecules[4]. However, this method often provides a low extraction yield due to the formation of irreversible precipitates at the organic-aqueous interface. In order to overcome this problem, we introduced a different approach to recover the proteins from the reversed micellar solution more effectively. This approach is the precipitation method in which a solid protein was recovered by the addition of a polar organic solvent. In the present study, acetone was used as the polar organic solvent.

The addition of a large amount (5-30 times in volume) of acetone to the reversed micellar solution provided 87-95 % recovery yields. Furthermore, the recovered RNase A retained a high enzymatic activity. The overall yield of the novel process was found larger in comparison with the conventional dilution method. Thus, this reversed micellar system was found to be effective for protein refolding even at high dosage of proteins. Our next objective is to extend the scope of this study to the direct refolding of inclusion bodies.

4. CONCLUSION

Solid denatured RNase A was completely solubilized into the AOT reversed micellar solution and was reactivated by the addition of the redox reagent glutathione. The novel refolding process is mainly divided into three steps: solubilization, renaturation and recovery steps. The renatured RNase A in reversed micelles was completely recovered by the addition of acetone. The benefits of this novel process are that high concentration of proteins can be treated and that the amount of the expensive redox reagent can be reduced due to the small water content in reversed micelles.

REFERENCE

1. A. J. Hagen, T. A. Hatton, and D. I. C. Wang, Biotechnol. Bioeng. 35(1990)955.
2. B. Batas, and J. B. Chaundhuri, Biotechnol. Bioeng. 50(1996) 16.
3. Y.Hashimoto,T. Ono, M. Goto, and T. A. Hatton, Biotechnol.Bioeng. 57(1998)620.
4. M. Goto, Y. Ishikawa, T. Ono, F. Nakashio, and T. A. Hatton, Biotechnol. Prog. 14(1998)729.

Bioseparation Engineering
I. Endo, T. Nagamune, S. Katoh and T. Yonemoto (Editors)

Development of Efficient Protein Refolding Systems Using Chaperonins

Jiro Kohda[a], Akihiko Kondo[b], Tadanaru Teshima[c] and Hideki Fukuda[a]

Division of [a]Molecular Science and [c]Bioscience, Graduate School of Science and Technology, and [b]Department of Chemical Science and Engineering, Faculty of Engineering, Kobe University, 1-1 Rokkodai-cho, Nada-ku, Kobe 657-8501, Japan

Application of chaperonins to *in vitro* protein refolding was investigated. *E. coli* chaperonin GroEL/ES and *Thermus thermophilus* holo-chaperonin (*T. thermophilus* holo-cpn) were used for construction of protein refolding systems. The systems based on immobilized chaperonin, fusion chaperonin and chaperonin in combination with ultrafiltration were investigated. Fusion chaperonin with the hexa-histidine affinity tag (GroEL-$(His)_6$) was efficiently purified by affinity column. The refolding activity of GroEL-$(His)_6$ was proved to be similar to that of native one. GroEL/ES was immobilized by covalent bond. In addition, GroEL-$(His)_6$ was immobilized on metal chelate resin by affinity interaction. These immobilized chaperonins retained sufficiently high refolding activities and were reused efficiently. Especially, immobilized GroEL-$(His)_6$ showed a high refolding activity. On the other hand, in ultrafiltration system, *T. thermophilus* holo-cpn, which forms a stable complex, was efficiently separated from refolded proteins and repeatedly used for protein refolding. The selection of the system is mainly dependent on the type of chaperonin most effective for the refolding of target proteins.

Key Words; Chaperonin, Protein Refolding, Affinity purification, Immobilization, Ultrafiltration

1. INTRODUCTION

Recently, it has been clear that molecular chaperones were constitutively expressed and play important roles in the synthesis of correctly folded proteins *in vivo* [1, 2]. In addition, molecular chaperones control wide range of cell function such as transcription, protein assembly and membrane translocation. Among molecular chaperones, chaperonins which play a central role in folding mechanism have been investigated actively. Bacterial chaperonins have cylindrical structure composed of two stacked 7-fold rotational symmetric rings of cpn60 subunit, and have co-chaperonin which is dome-shape structure composed of 7-fold rotational symmetric rings of cpn10 subunit. Bacterial chaperonins promote protein folding in an ATP-dependent manner.

The studies on applications of chaperonins to various fields in biotechnology have been started. Especially, application of chaperonins to *in vitro* protein refolding has been investigated actively. This is because recovery of active proteins from inclusion body by protein refolding is an important step in the production of many recombinant proteins. Although the recovery of active proteins from inclusion body is usually low, chaperonins are expected to significantly improve refolding yields.

The aim of this study was to construct efficient protein refolding systems based on chaperonins. For this purpose, it is important to optimize refolding conditions and to reuse chaperonins. Therefore, we studied protein refolding systems based on immobilized chaperonin (system 1), fusion chaperonin (system 2) and chaperonin in combination with ultrafiltration (system 3). Chaperonins from *E. coli* and thermophilic bacteria (*Thermus thermophilus*) were used. In the system 1, *E. coli* chaperonin GroEL/ES was immobilized by covalent bond. In the system 2, fusion chaperonin GroEL-$(His)_6$, which was constructed by fusing the hexa-histidine $(His)_6$ tag to carboxyl terminal of GroEL gene, was immobilized on metal chelate Cellulofine by affinity interaction. In the system 3, since *T. thermophilus* holo-cpn forms stable 21-mer complex, *T. thermophilus* holo-cpn was separated from refolded proteins by ultrafiltration.

2. MATERIALS AND METHODS

2.1. Overexpression and purification of chaperonin

Chaperonin GroEL/ES was affinity-purified on an immobilized casein column from lysates of *E. coli* DH5α cells harboring the plasmid pGroE, according to a method described elsewhere [3].

E. coli HMS174 (DE3) cells harboring the GroEL-$(His)_6$ expression plasmid, pGroEL-$(His)_6$ [4] and the *T. thermophilus* holo-cpn expression plasmid pRCC50 (Taguchi *et al.*, to be published elsewhere) were cultured at 37 ℃ in LB medium (10 g tryptone, 5 g yeast extract and 5 g NaCl per liter). Production of GroEL-$(His)_6$ and *T. thermophilus* holo-cpn was induced by the addition of isopropyl-β-thiogalactopyranoside (IPTG, final concentration 1 mM). After 3-h induction, cells were harvested by centrifugation at 5,000 ×*g* for 20 min at 4 ℃. Collected cells were resuspended in buffer A for GroEL-$(His)_6$ (20 mM TrisHCl buffer, pH 7.9, containing 0.5 M NaCl and 5 mM imidazole) and buffer B for *T. thermophilus* holo-cpn (50 mM TrisHCl buffer, pH 7.8, containing 10 mM $MgCl_2$, 20 mM KCl and 2 mM dithiothreitol (DTT)) and disrupted by ultrasonication. The disrupted cell suspension was centrifuged at 30,000 ×*g* for 30 min at 4 ℃. The supernatant containing GroEL-$(His)_6$ was loaded onto nickel chelate Cellulofine (Chisso Co.), which was prepared according to the manufacturer's instructions and was equilibrated with buffer A. After washing with 20 mM TrisHCl buffer, pH 7.9, containing 0.5 M NaCl and 60 mM imidazole, the bound proteins were eluted with 20 mM TrisHCl buffer, pH 7.9, containing 0.5 M NaCl and 1 M imidazole. The eluted fraction was dialyzed against 20 mM TrisHCl buffer, pH 7.5, containing 10 mM $MgCl_2$. To remove the most *E. coli* proteins, the supernatant containing *T. thermophilus* holo-cpn was heated at 75 ℃ for 30 min and centrifuged at 30,000 ×*g* for 30 min at 4 ℃.

2.2. Immobilization of chaperonins

As a solid support material, formyl-Cellulofine (Chisso Co.) was used, because it has a long spacer. GroEL/ES, bovine serum albumin (BSA; Nacalai Tesque) and glycine were coupled to formyl-Cellulofine in 0.1 M sodium phosphate buffer containing 0.5 M NaCl, pH 7.0 (coupling buffer) according to the manufacturer's instructions. In the coupling of GroEL/ES, 10 mM ATP (Oriental Yeast Co., Ltd.) was added to the coupling buffer. On the other hand, GroEL-$(His)_6$ was immobilized onto nickel chelate Cellulofine in buffer C (TrisHCl buffer, pH 7.5, containing 10 mM $MgCl_2$ and 2 mM ATP). After unimmobilized protein was washed away, the amounts of proteins in the washing solution were determined by the method of Bradford [5]. The amounts of immobilized chaperonins were estimated from the protein balance.

2.3. Measurement of enzyme activities

The activity of yeast enolase (47 kDa, dimer; Oriental Yeast Co., Ltd.) was measured at 25 ℃ using 2-Phosphoglyceric acid (Sigma Chemical Company) as described previously [6]. The activity of bovine DNase I (31 kDa, monomer; Worthington Biochemical Co.) was measured at 37 ℃ using calf thymus DNA (Worthington Biochemical Co.) as described previously [7].

2.4. Reactivation of thermally inactivated enolase by immobilized GroEL/ES and GroEL-(His)$_6$

Enolase was thermally inactivated at 53 ℃ in 50 mM TrisHCl buffer pH 7.8, containing 10 mM $MgCl_2$ and 20 mM KCl. Reactivation of enolase was performed at 37 ℃ by mixing the enzyme solution (30 μl) with immobilized GroEL/ES or GroEL-(His)$_6$ (gel volume 1 ml) suspended in buffer B. The final volume of reaction mixture was 2 ml, and the enzyme concentration was 0.2 μM. The reactivation mixture was gently mixed, and the activity of reactivated enolase was measured after separation of immobilized GroEL/ES or GroEL-(His)$_6$ by sedimentation. For the repeated utilization of immobilized chaperonins, the suspension was poured onto the column and washed with buffer B after the completion of each reactivation reaction. Then immobilized GroEL/ES or GroEL-(His)$_6$ was mixed with the thermally inactivated enolase again, and the next batch of reactivation was started.

2.5. Refolding of GdnHCl-denatured enolase and DNase I

Enolase and DNase I were mixed with an equal volume of 8 M Guanidine hydrochloride (GdnHCl; Wako Pure Chemical Industries) solution. The unfolded enzymes in 4 M GdnHCl were subsequently diluted by buffer C. The final concentration of GdnHCl was 60 mM. Specified amounts of chaperonins were added to buffer C where indicated. The final volume of the refolding mixture was 500 μl. The activities of refolded enzymes were measured as described above, and the refolding efficiency was evaluated as the specific activity of refolded enzymes relative to that of untreated enzymes.

2.6. Repeated utilization of *T. thermophilus* holo-cpn for enzyme refolding by the ultrafiltration system

Unfolded DNase I in 4 M GdnHCl (7.5 μl) were subsequently diluted into buffer B containing specified amounts of *T. thermophilus* holo-cpn and 2 mM ATP. The final volume of the refolding mixture was 500 μl. After incubation for 3 h at 30 ℃, *T. thermophilus* holo-cpn was separated from DNase I by ultrafiltration using YM100 (molecular cut-off, 100 kDa) membrane (Amicon Inc.). The refolding mixtures were filtrated until the volume was reduced to 250 μl. Then, 246.25 μl of buffer B containing 2 mM ATP and 3.75 μl of unfolded DNase I was added, and the refolding reaction of the next cycle was started. The refolding mixture was filtrated again until the volume was reduced to 250 μl. The enzyme activities of the filtrates were measured after each cycles. The refolding efficiency was evaluated as the percentage of refolded enzyme activity recovered in the filtrate relative to that of the native enzyme.

3. RESULTS AND DISCUSSIONS

3.1. Production of chaperonins

E. coli DH5α cells harboring the plasmid pGroE overproduced GroEL/ES by heat shock [3]. GroEL/ES was successfully purified with immobilized casein column because of affinity interaction between GroEL/ES and casein.

E. coli cells harboring the plasmid pGroEL-$(His)_6$ overproduced fusion chaperonin GroEL-$(His)_6$ in the soluble fraction. GroEL-$(His)_6$ was affinity purified by using nickel chelate Cellulofine [4]. GroEL-$(His)_6$ possessed the ATPase activity similar to that of GroEL [4]. From this result, the fusion of the $(His)_6$-tag to GroEL does not affect its ATPase activity. GroEL-$(His)_6$ was found to form a tetradecamer from the analysis of gel filtration chromatography [4]. Although the precise location of the carboxyl terminal of GroEL is unknown, above results indicate that the $(His)_6$-tag is located outside the chaperonin cylinder.

Since *T. thermophilus* holo-cpn is thermostable, *T. thermophilus* holo-cpn overproduced in *E. coli* was purified efficiently by heat treatment [8].

3.2. Reactivation of thermally inactivated enolase by free and immobilized GroEL/ES

Figure 1A shows the time course of the reactivation of thermally inactivated enolase by free GroEL/ES and GroEL/ES-Cellulofine. In the absence of additive, enolase activity decreased gradually. Although enolase was not reactivated in the presence of BSA-Cellulofine and glycine-Cellulofine, enolase activity was kept during incubation. Therefore, gel materials stabilize enolase. Enolase activity increased in the presence of free GroEL/ES and GroEL/ES-Cellulofine. However, its reactivation rate was slower than that of free GroEL/ES. This is due to denaturation of GroEL/ES during immobilization and large mass-transfer resistance of enolase into gel matrix.

Fig. 1B shows the effect of repeated cycles on the reactivation of thermally inactivated enolase. Although the activity of reactivated enolase decreased gradually, GroEL/ES-Cellulofine retained a sufficiently high activity during repeated cycles. This is because GroEL/ES is stable at the reactivation temperature. Immobilized GroEL/ES is effective for the refolding process because of its high reactivation activity and reusability.

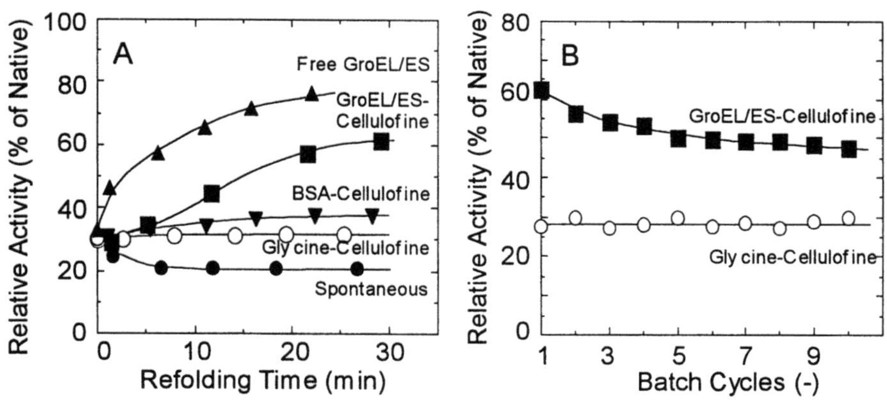

Fig. 1 (A) The time course of the reactivation of thermally inactivated enolase.
(B) The effect of repeated cycles on the reactivation of thermally inactivated enolase.

3.3. Reactivation of thermally inactivated enolase by free and immobilized GroEL-(His)$_6$

GroEL-(His)$_6$ promoted refolding of chemically denatured enolase to a similar extent as GroEL in the presence of ATP (Fig. 2A). Chaperonin inhibited the refolding of chemically denatured enolase in the absence of ATP, and delayed addition of ATP enhanced refolding. Since this result is similar to that of GroEL, GroEL-(His)$_6$ possesses the activity of protein refolding [3]. Since GroEL-(His)$_6$ is easily purified by affinity chromatography and possess high activity to promote protein refolding, immobilization of GroEL-(His)$_6$ on nickel chelate Cellulofine was examined. Immobilized GroEL-(His)$_6$ reactivated thermally inactivated enolase to a similar extent as free GroEL (Fig. 2B). Immobilized GroEL-(His)$_6$ possesses higher refolding activity than GroEL/ES-Cellulofine described above. Since GroEL-(His)$_6$ is immobilized by the affinity interaction between the (His)$_6$ tag and nickel chelate Cellulofine, a large percent of immobilized GroEL-(His)$_6$ remained its native structure. Although GroEL-(His)$_6$ was immobilized by noncovalent bond, immobilized GroEL-(His)$_6$ was repeatedly reactivate thermally inactivated enolase (Fig. 2C). Fusion chaperonin GroEL-(His)$_6$ is stable on metal chelate resin.

3.3. Repeated utilization of *T. thermophilus* holo-cpn for enzyme refolding by the ultrafiltration system

As shown in Fig. 3A, *T. thermophilus* holo-cpn refolded GdnHCl-denatured DNase I in the

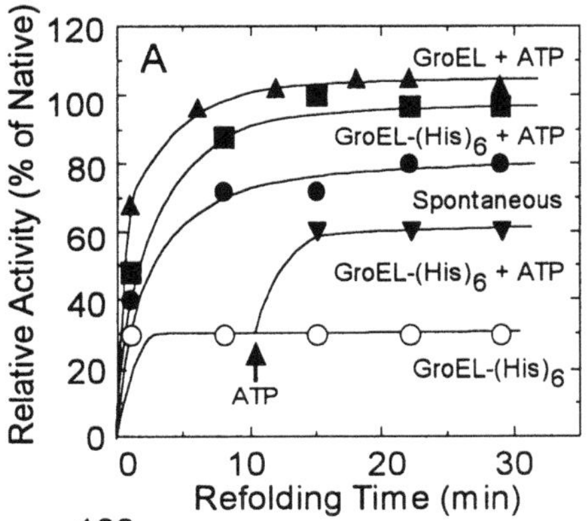

Fig. 2 (A) The time course of the refolding of GdnHCl-denatured enolase.
(B) The time course of the reactivation of thermally inactivated enolase.
(C) The effect of repeated cycles on the reactivation of thermally inactivated enolase.

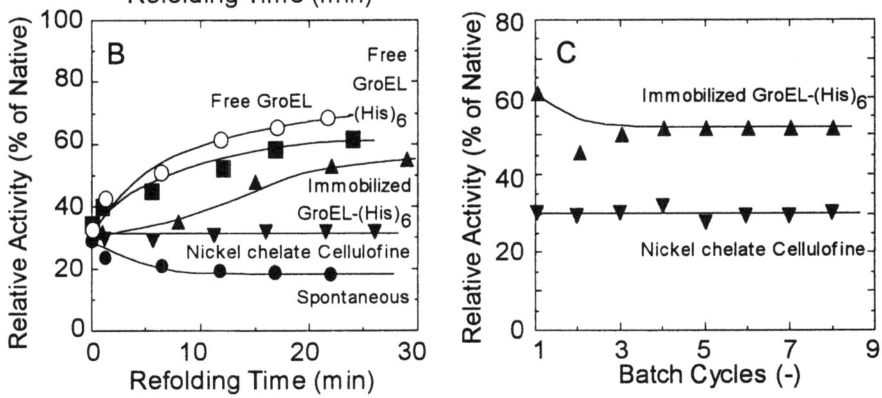

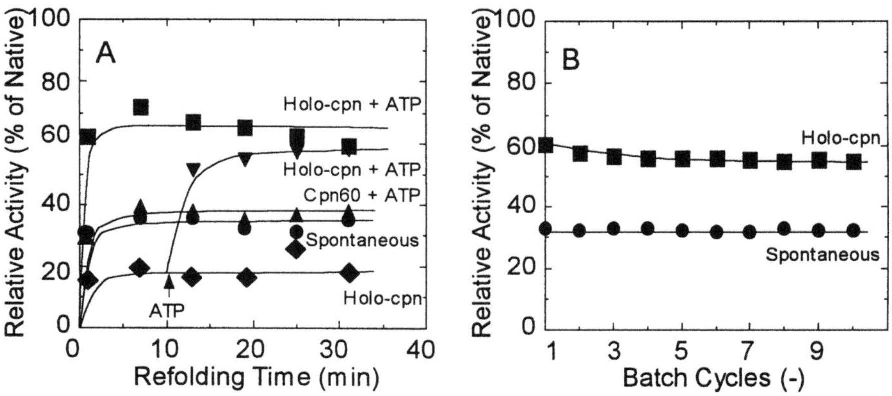

Fig. 3 (A) The time course of the refolding of GdnHCl-denatured DNase I.
(B) The effect of repeated cycles on the refolding of GdnHCl-denatured DNase I.

presence of ATP. The refolding of DNase I was inhibited in the absence of ATP, and DNase I activity was significantly increased by the addition of ATP. Since *T. thermophilus* holo-cpn exists as the stable 21-mer composed of cpn60 tetradecamer and cpn10 heptamer, the repeated utilization of holo-cpn was investigated by using the ultrafiltration system. *T. thermophilus* holo-cpn refolded GdnHCl-denatured DNase I repeatedly using the ultrafiltration system (Fig. 3B). Both *T. thermophilus* cpn60 and cpn10 were concentrated and separated from refolded DNase I by ultrafiltration [8]. This result indicates that *T. thermophilus* holo-cpn is stable during repeated refolding. Therefore, *T. thermophilus* holo-cpn is more stable than GroEL/ES [3]. Utilization of *T. thermophilus* holo-cpn in combination with the ultrafiltration system is effective for the refolding of enzymes which require co-chaperonin cpn10.

In conclusion, chaperonins are applicable for construction of the efficient protein refolding system. The selection of the system is mainly dependent on the type of chaperonin most effective for the refolding of target proteins.

REFERENCES

1. Gething, M. J., and Sambrook, J. (1992) *Nature* **355**, 33-45
2. Hartl, F. U. (1996) *Nature* **381**, 571-580
3. Ishii, Y., Teshima, T., Kondo, A., Murakami, K., Sonezaki, S., I-Ogawa, H., Kato, Y., and Fukuda, H. (1997) *Chem. Eng. J.* **65**, 151-157
4. Teshima, T., Mashimo, S., Kondo, A., and Fukuda, H. (1998) *J. Ferment. Bioeng.* **86**, 357-362
5. Bradford, M. M. (1976) *Anal. Biochem.* **72**, 248-254
6. Borders C. L. JR., Woodall, M. L., and George, A. L. JR., (1978) *Biochem. Biophys. Res. Commun.* **82**: 901-906
7. Thoru, P. (1972) *Proc. Natl. Acad. Sci. USA* **69**, 2224-2228
8. Teshima, T., Kohda, J., Kondo, A., Taguchi, H., Yohda, M., Endo, I., and Fukuda, H. (1998) *J. Ferment. Bioeng.* **85**, 564-570

Bioseparation Engineering
I. Endo, T. Nagamune, S. Katoh and T. Yonemoto (Editors)

Monitoring Structural Changes of Proteins on Solid Phase Using Surface Plasmon Resonance Sensor

Teruhisa Mannen[a, b], Satoshi Yamaguchi[a], Jun Honda[b], Shunjiro Sugimoto[b], Atsushi Kitayama[a], and Teruyuki Nagamune[a]

[a]Department of Chemistry & Biotechnology, the University of Tokyo, 7-3-1, Hongo, Bunkyo-ku, Tokyo, 113-8656, Japan.

[b]Bio-pharmaceuticals Development Center, Hoechst Marion Roussel Ltd., 1-3-2, Minamidai, Kawagoe, Saitama, 350-1165, Japan.

There are several techniques to detect the structural changes of proteins in solution. However it is difficult to apply them to the protein on solid phase. We show the possibility to monitor two kinds of signal changes for proteins immobilized on dextran resin by BIAcore biosensor based on surface plasmon resonance (SPR). Proteins with different properties were attached to sensor surfaces and various denatured states were induced by treatment with acidic or basic solutions. As a result, at least two different types of signal changes were detected real-time and these signal changes arose during and after the treatment with each solution that we denoted as *in situ* and *post* values, respectively. The *in situ* value seemed to have a strong correlation to the total charge state of the proteins which can be calculated theoretically, and the *post* value to the degree of structural changes of the proteins. This method is expected to be applied to various analyses and give us new information about the behavior of proteins on solid phase.

Key words
BIAcore, biosensor, surface plasmon resonance, immobilized protein, protein denaturation.

1. INTRODUCTION

Recently, various proteins of industrial or medical use are produced using heterologous gene expression system. These recombinant proteins, however, often form inactive, insoluble aggregate called inclusion body especially when they are expressed in *E. coli*. Therefore, protein refolding is considered to be one of the most important steps in the downstream process. Though some empirical strategies have been established for efficient protein refolding, they all have the common disadvantage of using huge tanks with large quantities of solutions due to a requirement of refolding condition under extremely low protein concentration, e. g.

10 μg/ml (1). To overcome this problem, a new refolding method capturing proteins on solid phase was proposed which realizes virtually infinite dilution of the protein. Stempfer *et al.* reported that electrostatically trapped α-glucosidase could be refolded with a high yield at a protein concentration of up to 5 mg/ml and refolding process of this protein could be monitored by measuring its enzymatic activity (2). This method, however, does not solve the problem about difficulties in optimizing refolding conditions and can only be applied to the proteins whose enzymatic activities are easily measurable. This is because conventional spectroscopic methods such as CD, UV, and fluorescence can not be easily applied to monitor refolding process of proteins on solid phase. General refolding method is still under research and at present, the optimum procedure has to be determined by trial and error.

When the optimum refolding condition is searched in liquid phase, it is difficult to re-use the same sample repeatedly because of the complicated re-purification of the protein. On the other hand, the immobilized protein on solid phase can be re-used repeatedly (3), thus a considerable improvement in the time-consuming optimization process is expected. By constructing an automatic system for buffer exchange operation and a monitoring system for refolding process of immobilized protein, high-throughput screening process of the optimum refolding condition can be established. Thus it is necessary to develop a new means for monitoring the structural changes of proteins on solid phase that is applicable more generally to any proteins.

2. EXPERIMENTAL METHOD

Surface plasmon resonance (SPR) sensor is now being utilized to detect the change of refractive index and, therefore, the mass density in accordance with molecular association or dissociation near the surface of thin metal layer. If a significant change of dielectric properties occurs in accordance with structural change of immobilized protein, we may be able to detect

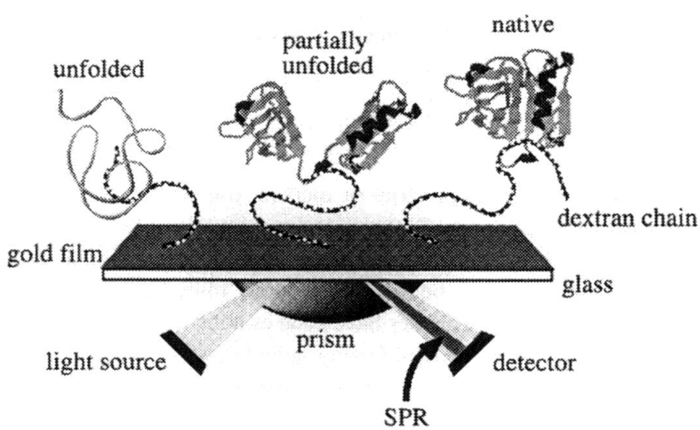

Figure 1. Can SPR see conformational change of immobilized protein on solid phase?

them as a change in refractive index because of Maxwell's equation, $\varepsilon = n^2$, where ε is a dielectric constant and n is a refractive index (Fig. 1). We, therefore, prepared sensor surfaces on which various proteins were immobilized covalently, and tried to detect signal changes using BIAcore 2000 (Biacore AB, Uppsala) when some denatured states were induced.

Carmody's wide range buffer series (the mixture of solution A: 0.2 M boric acid and 0.05 M citric acid and solution B: 0.1 M tertiary sodium phosphate at various ratios) were used as denaturing solutions. All operations to induce denatured states of immobilized proteins were done by pulse injections of each denaturing solution. Flow rate and the temperature of flow cell that included sensor surface were kept at 10 μl/min and 25 °C, respectively. Proteins on sensor surface were exposed to running buffer (0.1 M Tris-HCl, pH 7.6) before and after each pulse injection of denaturing solution.

3. RESULTS AND DISCUSSION

3.1. Evaluation of Sensorgram during Protein Denaturation

Time course of resonance signal from BIAcore is called a sensorgram. Signal a in Fig. 2 shows a typical sensorgram for acid denaturation of α-glucosidase. Two kinds of signal change depending on pH change were observed. One is the signal change X in the presence of acid and another is Y in the presence of running buffer. We denoted them *in situ* and *post* values, respectively, and investigated them in detail. The values just before finishing each pulse injection (*in situ* value) and just before starting next pulse injection (*post* value) were collected. Values from signal b (negative control) were subtracted from values from signal a. *In situ* and *post* values obtained were proportional to the amount of immobilized protein, indicating that signal changes derived solely from immobilized protein could be detected.

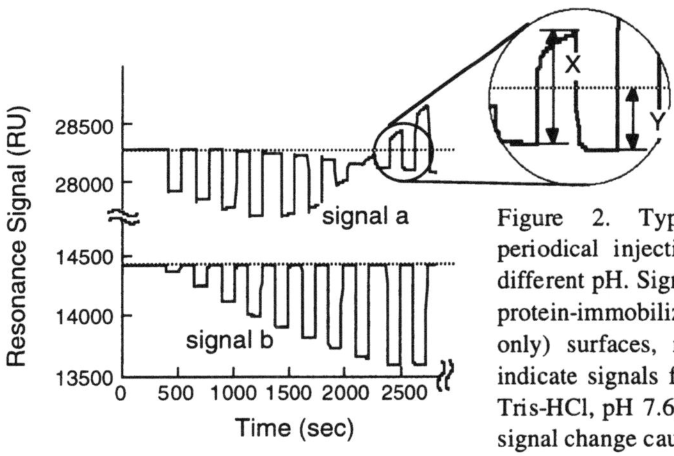

Figure 2. Typical sensorgram during periodical injection of acid solution with different pH. Signal a and b are signals from protein-immobilized and naked (CM-dextran only) surfaces, respectively. Broken lines indicate signals from running buffer (0.1 M Tris-HCl, pH 7.6). Each peak represents the signal change caused by each pulse injection (pH range from 7.6 to 1.9). Signal changes X and Y in the enlarged figure are denoted as *in situ* and *post* values, respectively.

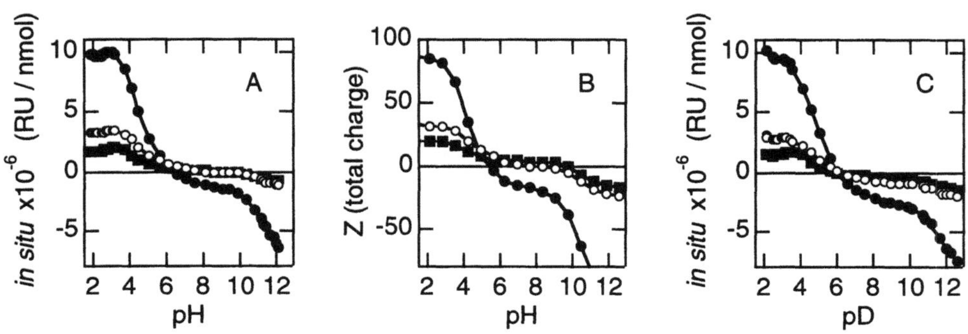

Figure 3. Analysis of *in situ* values of immobilized proteins. *In situ* values of three proteins as a function of pH (A), pD (C), and charge states of three proteins calclulated theoretically (B) are shown. *In situ* values are displayed as the changes of resonance unit per nmol molecules. ●, α-glucosidase; ■, α-chymotrypsinogen; and ○, myoglobin.

3.2. Signal Change during Denaturation Treatment: *in situ* Value.

Fig. 3A depicts pH dependence of *in situ* values of three kinds of proteins. This result was obtained by combining the results of successive acid (from pH 7.6 to pH 1.9) and alkali (from pH 7.6 to pH 12.2) pulse injections. Though pH dependencies of the values were different from one another, all showed positive values at acidic region and negative values at alkaline region. This feature was independent of the properties of secondary structures of each protein (myoglobin, all α helical; α-chymotrypsinogen, all β sheet; α-glucosidase, mixture of the two). We speculated that the cause for this behavior of signal changes accompanied by protonation and deprotonation of proteins depending on the pH of solution as: (a) the change of charge state; (b) the change of mass; or (c) the structural difference between the acid and alkali-induced denatured states. Based on these speculations, the changes of theoretical charge state of each protein were calculated according to the equation:

$$Z = -\sum_i \frac{ai\,Ki}{[H] + Ki} + \sum_j \frac{bj\,[H]}{[H] + Kj} \tag{1}$$

where Z is total charge of protein, [H] is proton concentration, Ki is the Ka (equilibrium constant, 10^{-pKa}) of acidic residue, ai is its number per protein molecule, Kj is the Ka of basic residue, and bj is its number per protein molecule. The result of calculation on each protein is shown in Fig. 3B, and they seemed to have a strong correlation with *in situ* values. To verify this and speculation (b), the same experiment was conducted using heavy water — same charge but different mass (Fig. 3C). The result was almost the same as that from light water, indicating that isotope effect on mass density was too small to be detected by BIAcore. As for speculation (c), we know that the acid and alkali denatured states can not be clearly discriminated by CD or fluorescence spectra in liquid phase because both denatured states resemble each other in secondary and tertiary structures and this is also expected for both states of denatured proteins on solid phase. Thus, it can be concluded that there must be a strong correlation between the total charge of protein and the refractive index, and BIAcore can monitor the charge state of immobilized proteins.

3.3. Signal Change after Denaturation Treatment: *post* Value

Using myoglobin (holo-Mb) and apomyoglobin (apo-Mb), successive acid and alkali pulse injections were performed, and *post* values were collected. As shown in Fig. 4, in contrast to *in situ* values, *post* values of both holo-Mb and apo-Mb decreased sharply at extreme acidic and alkaline conditions. It is noteworthy that the *post* value showed the same behavior during both acid and alkali pulse injections, which suggested that this value represents the structural changes of protein on solid phase.

In the case of acid denaturation of holo-Mb, the decrease occurred in two steps: the first step took place in the pH range from 6.0 to 4.0 and the second from 3.0 to 2.0, and between these pH ranges, *post* value increased. This behavior also depended on the ionic strength of pulse injected solution. For example, pH range of the first-step decrease (from 6.0 to 4.0) at high ionic strength condition shifted to higher pH range at low ionic strength condition (from 7.0 to 5.0). On the contrary, the two-step decrease in *post* value was not observed for apo-Mb.

Differences between holo-Mb and apo-Mb are: (a) presence of heme molecule; and (b) only apo-Mb has a highly plastic molten globule-like structure (4). Taking (a) into consideration, the first-step decrease observed only in holo-Mb was expected to represent the dissociation of heme molecule in acidic condition. However, from UV-measurement, the dissociation of heme easily occurred rather at higher ionic strength condition (data not shown). Thus this decrease was more likely to be derived from the difference (b), namely, it probably represents the structural change of holo-Mb from native to partially unfolded structure, that is similar to the structure of apo-Mb. In fact, proteins are less stable at lower ionic strength condition, and it seems that the collapse of the structure at higher pH at lower ionic strength resulted in the pH-shift of the first-step decrease. The decrease of signal change means the decrease of refractive index, and hence the decrease of dielectric constant around the solid surface. Therefore

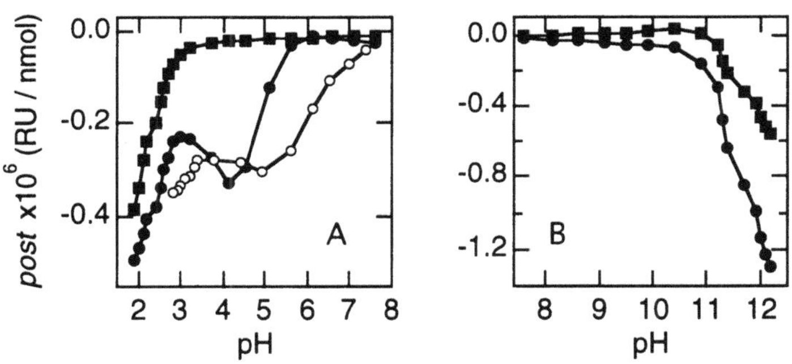

Figure 4. *Post* values of holo-Mb and apo-Mb induced by successive acid (A) and alkali (B) pulse injections. Carmody's buffer series (● and ■; pH range from 7.6 to 1.9 and from 7.6 to 12.2) and 10-fold diluted (○; pH range from 7.4 to 2.8) were used for denaturation. ● and ○, holo-Mb; ■, apo-Mb.

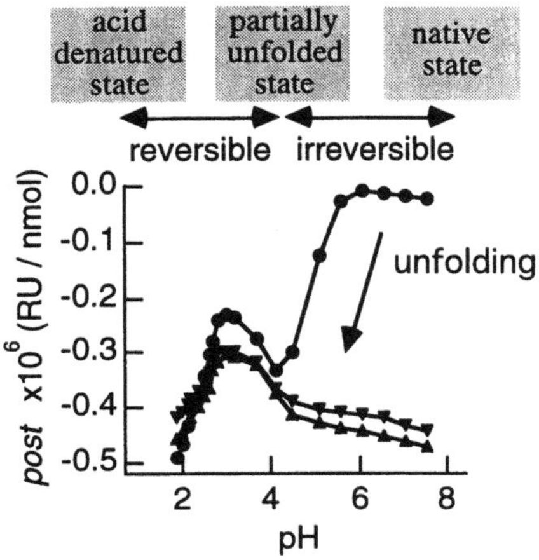

Figure 5. *Post* values of repeated successive pulse injections. The one run of this operation was performed repeatedly with 30-minute intervals using same holo-Mb. Carmody's buffer series was used for denaturation. ●, first run; ▲, second run; and ▼, third run. Probable states of the protein are indicated above the figure.

the *post* value is indicative of the denaturation of myoglobin resulting in the decrease of dielectric constant of dextran layer that bore myoglobin. This is consistent with the fact that the denaturation of myoglobin causes the disappearance of helices which has significant dipole moments.

To verify these possibilities, repeated successive acid pulse injections were performed using the same sample of protein on solid phase. Interestingly, quite a different pattern of *post* value was obtained in the second and the third runs as compared with the pattern in the first run (Fig. 5). In addition, the values at extreme acidic pH of each run tended to converge to the same value. Note that the data from the second run of holo-Mb was clearly different from those of the first run of apo-Mb (Fig. 4A, ■ and Fig. 5, ▲). This result also indicates that the first-step decrease of *post* value only seen in holo-Mb does not represent the dissociation of heme itself. Thus, it is suggested that *post* value represents the structural changes of myoglobin. What kind of structural parameter of protein directly influences the changes in *post* value is still unknown, but further studies on other proteins will reveal the theoretical aspect of signal changes accompanied by protein denaturation.

REFERENCES

(1) Rudolph, R.; Lilie, H., FASEB J.,10 (1996) 49-56. *In vitro* folding of inclusion body proteins.

(2) Stempfer, G.; Höll-Neugebauer, B.; Rudolph, R., Nat. Biotechnol.,14 (1996) 329-34. Improved refolding of an immobilized fusion protein.

(3) Hayashi, T.; Matsubara, M.; Nohara, D.; Kojima, S.; Miura, K.; Sakai, T., FEBS Lett.,350 (1994) 109-12. Renaturation of the mature subtilisin BPN' immobilized on agarose beads.

(4) Lin, L.; Pinker, R. J.; Forde, K.; Rose, G. D.; Kallenbach, N. R., Nat. Struct. Biol.,1 (1994) 447-52. Molten globular characteristics of the native state of apomyoglobin.

Chapter 3

Partitioning and Extraction

Bioseparation Engineering
I. Endo, T. Nagamune, S. Katoh and T. Yonemoto (Editors)

Recent advances in reversed micellar techniques for bioseparation

S. Furusaki[a], S. Ichikawa[b] and M. Goto[a]

[a]Department of Chemical Systems & Engineering, Kyushu University,
Fukuoka 812-8581, Japan

[b]Institute of Applied Biochemistry, University of Tsukuba,
Tsukuba 305-8572, Japan

Biocompatible systems of reversed micelles are required for the application to the preparation of food additives or medicinal substances. Use of soybean lecithin or phosphatidylcholine as a surfactant, and ethyl oleate, ethyl linoleate or ethyl caproate as a solvent gave satisfactory systems for this purpose. Oleic acid or cholesterol can be used as a cosolvent. Characterization of the micelles using small angle X-ray scattering (SAXS) is presented.

Extraction of DNA using reversed micelles of positively charged surfactants is possible. The effect of carbon number of the surfactants was studied, and distearyl quaternary ammonium chloride was most effective. Alcohol was used as a cosolvent and 1-octanol gave the best result. The circular dichroism (CD) spectrum of the extracted DNA was the same as that in an aqueous solution.

1. INTRODUCTION

Studies of the application of reversed micelles have been carried out particularly for separation of proteins and for enzyme reactions in organic media. In order to use the reversed micelles for preparation of biological products to be used as food additives or medicinal substances, the system forming the micelles should be biologically compatible. Recently, there have been substantial developments in the preparation of the biocompatible reversed-micellar systems. Also several applications of reversed micelles have been developed such as refolding of proteins and extraction of DNA's and nucleotides/nucleosides. Here, recent results on the biocompatible systems and also on the extraction of DNA are presented.

2. BIOCOMPATIBLE SYSTEM

Several biocompatible reversed micellar systems have been proposed (1-5). However, the solvents of their systems are not strictly biocompatible except hexane, although their surfactants were mostly biocompatible. Hexane can be used as a solvent to extract vegetable oils, but it is still uncertain whether hexane is totally biocompatible.

Reversed micelles solely composed by natural products can be constructed if we use natural surfactants and solvents. The formation of reversed micelles can be seen by water content in the organic phase containing surfactants. Figure 1 shows the water content with respect to surfactant concentration in organic phase. Above the critical micelle concentration (cmc), the water content increased significantly. Thus, the system with lecithin as a surfactant or ethyl caproate and ethyl oleate as a solvent was found to form reversed micelles. The system with hexane as a solvent, which forms reversed micelles as well, is also shown in the figure as a comparison.

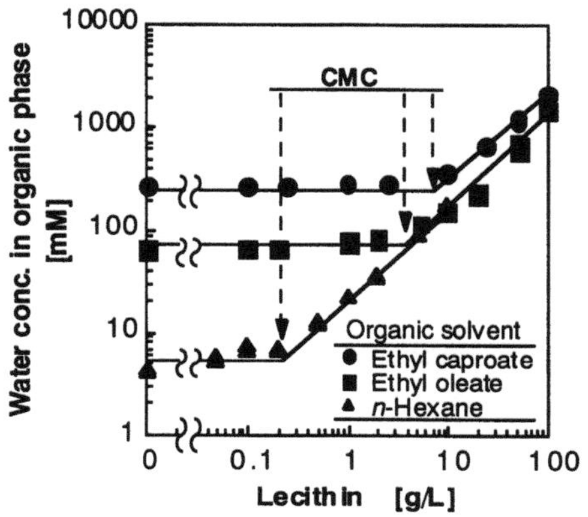

Figure 1. Solubilization of water into organic phase by the formation of lecithin reversed micelles.

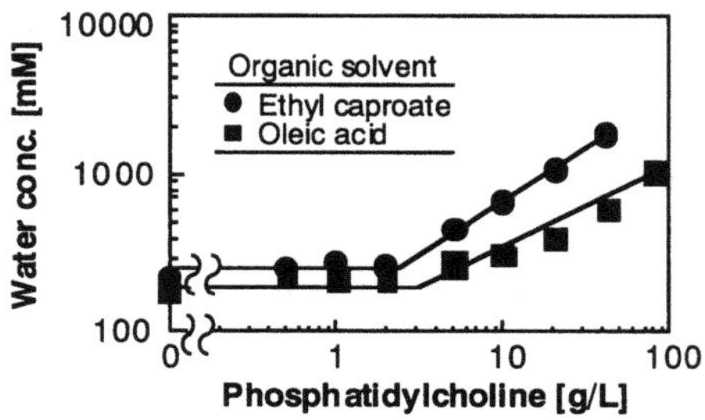

Figure 2. Solubilization of water into organic phase by the formation of phosphatidylcholine reversed micelles.

Since soybean lecithin is a mixture of several components, we tried the use of phosphatidylcholine (PC), which is one of the main components, as a surfactant. It was found that the water content of the organic phase increased as shown in Figure 2 when we used ethyl caproate or oleic acid as a solvent. Thus, reversed micelles were formed in these systems. After several combinations of surfactants and organic solvents, the following combinations were found to form reversed micelles; lecithin-ethyl caproate, lecithin-ethyl oleate, lecithin-ethyl linoleate, PC-ethyl caproate, PC-oleic acid. Addition of oleic acid or cholesterol to the PC-ethyl oleate system increased the water content in the organic phase remarkably. Therefore, we can say that oleic acid or cholesterol can assist formation of reversed micelles for the biocompatible systems.

Small angle X-ray scattering (SAXS) was measured to know the shape and size of the obtained reversed micelles. The Kratky plot showed that the micelles were spherical. The

Guinier plot of the SAX data gave the information that the diameter of micelles formed by the partitioning method was almost identical for different lecithin concentration in the system (Fig. 3). The water content in the organic phase increased. This means that the number of the micelles increased with the increase of the lecithin concentration. Instead, when the micelles were made by the injection method, the diameter changed with the water concentration in the organic phase (data not shown).

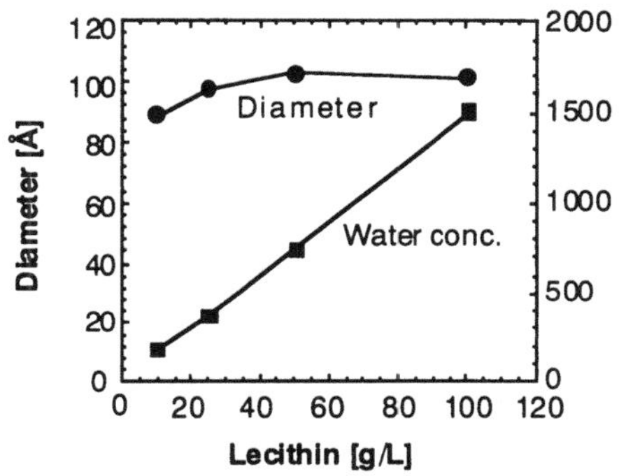

Figure 3. Effect of lecithin concentration on the micellar diameter and the water concentration in the reversed-micellar system formed by the partitioning method.

Extraction of cytochrome *c* was carried out to see whether the biocompatible system can be used for extraction of proteins. The visible light absorption (350 - 450 nm) of the organic phase representing the heme group after the contact was about the same as that of the original aqueous solution. Thus, the data shows that the extraction of cytochrome *c* was possible by this biocompatible system (6).

3. EXTRACTION OF DNA BY REVERSED MICELLES

Extraction of DNA was investigated with the reversed micellar system. Since DNA has negative charges due to its phosphate group, the surfactants with positive charges were applied for the extraction. The DNA studied was from salmon testes with molecular weight of *ca*. 6,500,000. It was purified by the method stated elsewhere (7). Extraction was carried out in a test tube at 298 K by using equal volumes (5 ml) of the aqueous and organic phases. Among several surfactants studied, di-stearyl quaternary ammonium chloride ($2C_{18}QAC$) with isooctane as the solvent and 1-octanol as a cosolvent gave a transfer yield of 100 % extraction at pH 7 - 8. The effect of the surfactant concentration on the fraction of DNA transferred to the organic phase is shown in Fig. 4. The CD spectrum of the extracted DNA did not change from the original CD spectrum in the aqueous phase (data not shown).

Thus, the double helical structure of DNA seems to be preserved during the course of extraction.

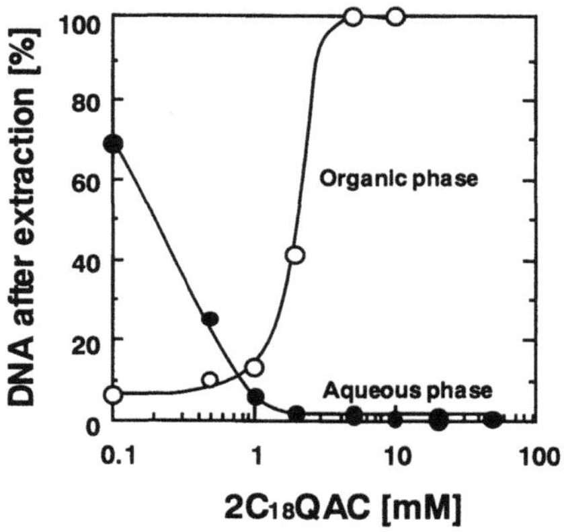

Figure 4. Effect of surfactant concentration ($2C_{18}QAC$) on the percentage extraction of DNA from aqueous phase into organic phase. Extraction was performed under following conditions: 10 mM Tris-HCl buffer, pH 8, DNA 50 μg/ml, salt-free and octanol 5 %(v/v).

REFERENCES

1. A. Ohshima, H. Narita and M. Kito, J. Biochem., **93** (1983) 1421.
2. P. Walde, A. M.Giuliani, C. A. Boicelli and P. L. Luisi, Chemistry & Physics of Lipids, **53** (1990) 265.
3. Y. Yamada, C. Kasai, R. Kuboi and I. Komasawa, Kagaku Kogaku Ronbunshu, **20** (1994) 54.
4. M. Vasudevan, K. Tahan and M. Wiencek, Biotechnol. Bioeng., **46** (1995) 99.
5. K. Naoe, M. Nishino, T. Ohsa, M. Kawagoe and M. Imai, J. Chem. Technol. Biotechnol., **74** (1999) 221.
6. S. Sugiura, S. Ichikawa, M. Nakajima, Y. Sano, M. Seki and S. Furusaki, Chem. Eng. Symp. Ser. No. 63, 131 - 138, Soc. Chem. Eng. Japan (1998).
7. M. Goto, T. Ono, A. Horiuchi and S. Furusaki, J. Chem. Eng. Japan, **32** (1999) 123.

ACKNOWLEDGMENT

The authors thank A. Momota, Kyushu University, for taking the experimental data of the DNA extraction.

Bioseparation Engineering
I. Endo, T. Nagamune, S. Katoh and T. Yonemoto (Editors)

A Novel Method of Determining the Aggregation Behavior of Microemulsion Droplets

Wen-Yih Chen, Chih-Sheng Kuo, and Der-Zen Liu

Department of Chemical Engineering, National Central University Chung-Li, Taiwan 320

The study proposes a novel method of obtaining the information of second virial coefficient of interactions between microemulsion droplets by microcalorimetry. By employing a high sensitivity isothermal titration microcalorimetry (ITC) to measure the dilution heat of microemulsions solution, the information of second virial coefficient of the interactions between the microemulsion droplets can be derived with the number density of microemulsion solution. The derivation is based on a hard-sphere interaction potential assumption. The variance of the derived second virial coefficients were verified by the percolation behavior of different reverse micelles solutions of dioctyl sulfosuccinate sodium salt (AOT) in decane with or without solutes. The well-correlated data between the percolation temperatures and the second virial coefficients between the droplets prove the feasibility of ITC dilution measurement as a tool for determining the behavior of the non-ideal collisions between microemulsion.

1. INTRODUCTION

Microemulsions are thermodynamically stable mixtures of water, oil, and surfactant(s) that exhibit a rich phase behavior. The interactions between microemulsion droplets account for the concerns of the dispersity in various applications, such as a drug carrier, cosmetic applications, and as a bioreactor. For the discussions of the interaction between microemulsion droplets, similar to the McMillan-Mayer theory [1] for molecular solution, the solvent was treated a continuum because the droplets consist of many molecules and much larger than the solvent molecules. Although the droplets are in dynamic equilibrium between the dispersed phase and surfactants, the interaction potential appears to exhibit a distinct hard-core part with a radius. Basically, the interaction forces between droplets involve the attractive van der Waals between surfactants interface and, for ionic surfactant, the electrostatic repulsive force. And, for the liquid film, such as the surface of the microemulsion, the replusive hydration and the entropic forces in a short range distance have contribution to the pair potential and is an energy barrier [2]. However, the direct measurement of the short range forces between the droplets has not been achieved successfully, therefore, a thermodynamic aspect of the energy potential or a description of the deviation from elastic collision, such as the second virial coefficient, is needed for describing different droplets interaction. The developments of the literature regarding the interactions between microemulsion droplets have been reviewed by Koper et al.[3].

The second virial coefficient has been experimentally determined by measuring of the osmotic pressure [4] and diffusion coefficient as function of volume fraction [5]. And the magnitude (and sign) of the second virial coefficient was correlated with the aggregation phenomenon of microemuslion. This letter presents a novel idea of using the isothermal titration microcalorimetry to measure the dilution heat of the microemulsion solution and, with the statistic thermodynamics derivation and a hard-sphere assumption, the information of second virial coefficient of the interaction between droplets can be obtained. Furthermore, the results of the ITC method was verified by the variance of the second virial coefficients with the percolation temperatures of AOT reverse micelles with or without solutes. The formation of percolating clusters by the reverse micelles can be attributed by the non-elastic collisions between the droplets, and can be monitored by the electric conductivity change of the solution by changing the microemulsion solution temperature.

In this study, conductivity measurement is designed to detect the conductivity percolation temperature of various reverse micelle systems with or without solute. Whereas the microcalorimetric study performed herein provide interaction potential between droplets, thereby allowing the microcalorimetric results to be explained in interaction potential aspect.

2. RESULTS AND DISCUSSION

The percolation temperature is defined as the peak temperature of the rate of change in the log value of the conductivity verses the temperature plot. The effects of the various solutes on the percolating cluster formation of reverse micelles have been extensively discussed and reported [6-9].

With the discussion above, the dilution heat of the solution verses the concentration of reverse micelles was measured by ITC, combing with the virial equation for non-ideal behavior of microemulsion solution, the present study proposes the following derivations and discussions for the derivation of second virial coefficient from dilution heat.

In general, the dilution heats of the reverse micelles solution with the water volume fraction can be fitted by a function of second order polynomial as Eq. (1)

$$\frac{d\dfrac{q}{Nk_BT}}{d\Phi_W} \cong \frac{d\dfrac{E}{Nk_BT}}{d\Phi_W} = b_2 + b_3\Phi_W + b_4\Phi_W{}^2 \qquad (1)$$

where q : heat of dilution (mJ)

E : internal energy (mJ)

N : number of particles (number of reverse micelles droplets in the solution)

k_B: Boltzman constant

T : absolute temperature (K)

Φ_W: water volume fraction

For an open liquid system, there is no pressure change and the dilution volume is neglected comparing with the total system solution volume, the dilution heat observed then is equal to the internal energy change of the system. Furthermore, the internal energy changes of the system with the number density of the reverse micelles in the solution can be expressed by the virial equation as Eq. (2)

$$\frac{E}{Nk_BT} = \frac{3}{2} - T\sum_{i=1}^{\infty}\frac{1}{i}\frac{dB_{i+1}}{dT}\rho^i \tag{2}$$

where $\rho=\Phi_W/V_d$ (V_d is the volume of the reverse micelle droplet). Comparing Eq.(1) with Eq.(2) reveals that the coefficient of the polynomial fitting of the experimental dilution heat data can be affiliated with the virial coefficients of Eq.(2) by the following Eq.(3)

$$V_d \times b_2 = -T\frac{dB_2}{dT}, \quad V_d^2 \times b_3 = -T\frac{dB_3}{dT} \tag{3}$$

From statistic thermodynamics, the second virial coefficient can be represented by the interaction potential energy function U(r) as Eq.(4)

$$B_2(T) = -2\pi\int_0^{\infty}[\exp(-\frac{U(r)}{k_BT}) - 1]r^2dr \tag{4}$$

If a square-well attractive potential energy function U(r) is selected and plugged into Eq.(4), the Eq.(3) can be declared as following:

$$V_d \times b_2 = -T\frac{dB_2}{dT} = -B_0(\lambda^3 - 1)e^{\varepsilon/k_BT}(\varepsilon/k_BT) < 0 \tag{5}$$

where $B_0=(16/3)\pi R_{HS}^3$

$\lambda=(2R_{HS}+\sigma)/2R_{HS}$

and σ, ε denotes the width and depth of the square-well attractive potential function, respectively.

Analyzing Eq.(5) reveals, qualitatively, that a negative b_2 value indicates an attractive interaction and a more negative value of b_2 (higher value of ε and σ) declare that a non-elastic collision is more likely to happen between the reverse micelle droplets. Considering the percolation behavior from the perspective of interaction potential, the more attractive interaction potential (higher value of ε and σ) would results in a system of reverse micelles solution easier to form percolation (lower T_P value). Therefore, if two-body interaction is considered only, a larger negative value of second virial coefficient indicates a lower T_p value of the reverse micelle solution.

The above attempt was demonstrated as following: The dilution heat of the reverse micelle solution with various water volume fraction of (AOT/Water/n-decane with 30mM $CuCl_2$) was measured and the heat generated with the water volume fraction were fitted by a second order polynomial fitting. The b_2 values of the fitting equation (Eqn.(1)) were obtained for various systems and were listed in Table 1 with the T_P values from the conductivity measurements. In summary, b_2 value from a thermodynamics aspect derivation of a dilution can be qualitatively well correlated with the T_p from conductivity percolation measurement. Above results not only can be used to describe the conductivity percolation from a thermodynamic perspective but also suggest that a simple dilution heat can be used to inquire the interaction potential information between reverse micelle or, in general, between colloids. The results can also provide, possibly, more details discussions of the colloid interaction mechanism.

3. CONCLUSIONS

This study, we have developed a microcalorimetry method of obtaining the second virial coefficient of the interaction potential between microemulsion droplets, and the validity of the coefficient were examined by the temperature of percolating cluster forming of various reverse micelles systems with different solutes. This novel method should be able to serve as one of the method of determining the stability of the microemuslion and also helpful of understanding the interaction mechanism of microemulsion droplets systems.

Table 1
The b_2 value and the conductivity percolation temperatures of various reverse micelle solution systems.

Reverse Micelle Solution	b_2 value	Tp(°C)
AOT/Decane/[$CuCl_2$]=30mM	-2.178×10^{-20}	30.3
AOT/Decane/[NaCl]=30mM	-5.202×10^{-20}	24.3
AOT/Decane/Water	-1.164×10^{-19}	14.1
AOT/Decane/[Trp]=20[RM]	-1.304×10^{-19}	13.2

REFERENCES

1. T.L.Hill, An Introduction to Statistical Thermodynamics, Dover Publishers Inc, New York, 1986.
2. J. Israelachvilli, Intramolecular and Surface Forces, Academic Press Ltd., London, 1992.

3. G.L.M. Koper, W.F.C. Sager, J. Smeets, and D. Bedeaux, J. Phys. Chem. 99 (1995) 13291.

4. D. Stauffer, Introduction to Percolation Theory, Taylor and Francis Inc. London, 1985.

5. G. Cassin, S. Illy, and M.P. Pileni, Chem. Phys. Lett. 221 (1994) 205.

6. A.S. Bommarius, J.F.D. Holzwarth, C. Wang, and T.A. Hatton, J. Phys. Chem. 94 (1992) 7232.

7. M.P. Pileni, (eds.), Structure and Reactivity in Reverse Micelles, Elsevier, Amsterdam, 1989.

8. M.A. Rodgers, and P.C. Lee, J. Phys. Chem. 88 (1984) 3480.

9. M.P. Pileni, T. Zemb, and C. Petit, Chem. Phys. Lett. 118 (1985) 414.

Bioseparation Engineering
I. Endo, T. Nagamune, S. Katoh and T. Yonemoto (Editors)

Preparation of Temperature-Sensitive Antibody Fragments

Masamichi Kamihira and Shinji Iijima

Department of Biotechnology, Graduate School of Engineering, Nagoya University
Furo-cho, Chikusa-ku, Nagoya 464-8603, Japan

We designed a temperature-sensitive single chain antibody fragment in which the antigen-binding activity was drastically altered by a temperature-shift between 4 and 37°C. An oligonucleotide corresponding to the temperature-sensitive helix-coil transition peptide, -(-Glu Ala Ala Ala Lys-)-, was introduced between VH and VL genes of scFv against the simian virus 40 large T antigen. The antigen-binding activity of the temperature-sensitive scFv produced by *Escherichia coli* cells showed a maximum decrease of 1/36 at 4°C compared with that at 37°C. Binding activity was controlled by the NaCl concentration, as well as by the temperature shift. By using a column that immobilized the scFv, the antigen was purified in response to elution by the temperature-shift.

1. INTRODUCTION

A high specificity and strong affinity characterize antigen-antibody binding. This has provided researchers with a strong tool for analysis and purification of trace amounts of an antigen. However, this strong affinity binding also makes it difficult to dissociate an antigen from its corresponding antibody. This is one of the problems associated with immuno-affinity chromatography.[1)]

On the other hand, progress in genetic engineering has enabled isolation of a gene corresponding to variable regions of an antibody from either hybridoma or spleen cells. It has also enabled expression of *Escherichia coli* as a single chain antibody fragment (scFv), in which antigen-

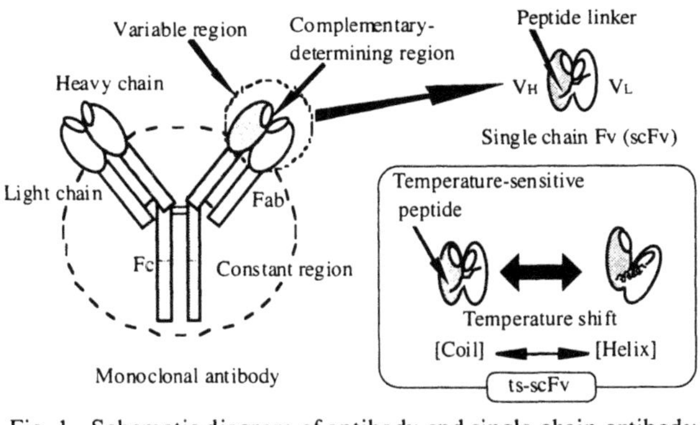

Fig. 1. Schematic diagram of antibody and single chain antibody fragment (scFv)

binding sites, including variable regions of heavy and light chains of the antibody, are linked through a linker peptide (Fig. 1).[2-5] Previous research has reported that the length of the linker peptide affects antigen-binding activity of scFv. In most cases, a flexible peptide sequence, -(-Gly Gly Gly Gly Ser-)$_3$- was used as the linker peptide. The scFv technique has potential applications in antibody engineering, such as genetic design of an antibody fragment.[3,4]

In the present study, a temperature-sensitive helix-coil transition peptide was genetically introduced into scFv as a linker peptide in order to control antigen-binding activity by altering the temperature (Fig. 1). The constructed scFv gene was expressed in *E. coli* cells, and clones that produce temperature-sensitive scFv (ts-scFv) were selected. The ts-scFv was immobilized on Sepharose support and used for temperature-dependent affinity purification of the antigen.

2. MATERIALS AND METHODS

2.1. Preparation of temperature-sensitive scFv

An scFv gene of a monoclonal antibody against the simian virus 40 large T antigen was prepared from a mouse hybridoma cell line, PAb1400, obtained from RIKEN cell bank (RCB0026) using a kit according to Pharmacia Biotech AB. The isolated scFv gene was subcloned into sequence vectors and then sequenced by a DNA sequencer. A set of PCR primers, which include restriction enzyme sites to the ends of the genes, was designed to amplify the VH and VL genes. The linker oligonucleotides corresponding to the temperature-sensitive helix-coil transition peptide, -(-Glu Ala Ala Ala Lys-)$_n$-, previously reported by Merutka *et al.*,[6,7] were synthesized by PCR using 5′-TTTGCTAGCCTTTGCTGCGGCTTCCTTTGCTGCGGCTTCCTTTGCTGCGGCTT CTTTAGCGGCCGCCTCTTTGGCAGCTGCCTCGGCGCCATACTCTAAAGT-3′ as a template, and 5′-ACTTTAGAGTATGGCGCCGAG-3′ and 5′-TTTGCTAGCCTTTGCT

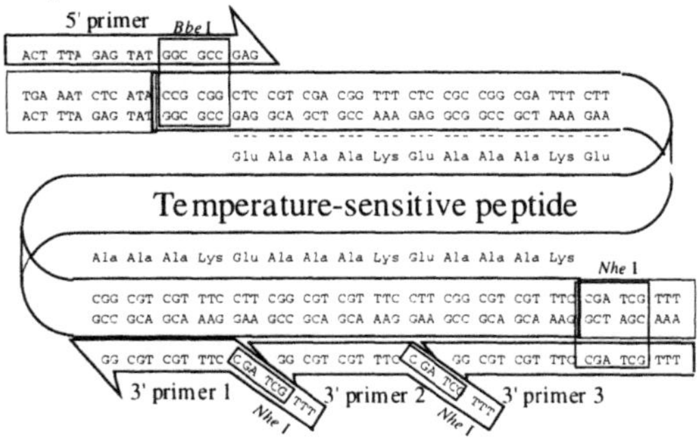

Fig. 2. Synthesis of oligonucleotides for temperature-sensitive linker peptide

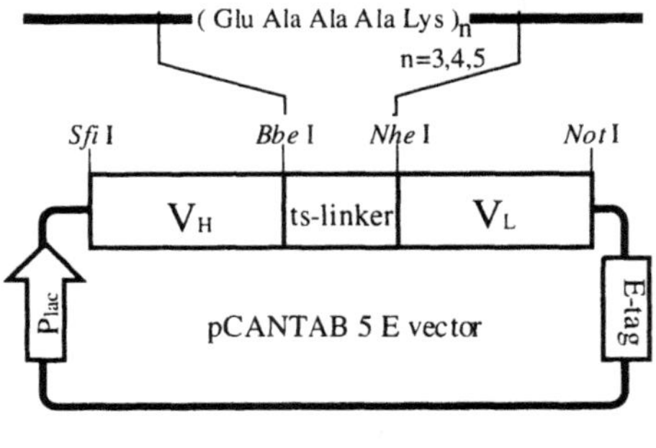

Fig. 3. Plasmid construction for expression of temperature sensitive scFv

GCGG-3′ as primers (Fig. 2). The 3′-primer was capable of annealing to three different sites in the template. Thus, three linker oligonucleotides differing in terms of their length -(-Glu Ala Ala Ala Lys-)$_n$- [n=3-5] were synthesized after PCR amplification. Thereafter, the respective linker oligonucleotides were used as the template for second PCR using Taq polymerase in order to add mutation to the linker sequence. After restriction enzyme digestion of V_H, V_L and linker genes, the oligonucleotides were ligated and inserted into a pCANTAB5E vector (Pharmacia Biotech AB) (Fig. 3). *E. coli* HB2151 cells were transformed using the vector plasmids to produce scFvs, and the *E. coli* clones were selected on the basis of drastic changes in the antigen-binding activity between 4 and 37°C.

2.2. Production of scFv

Overnight culture of *E. coli* HB2151 harboring the plasmid was seeded onto fresh SB-AG medium (35 g polypepton, 20 g yeast extract, 5 g NaCl, 20 g glucose and 100 mg ampicillin per liter, pH7.5). The culture broth was incubated for 1 h at 30°C. Following removal of the medium by centrifugation, the cells were resuspended in SB-AI induction medium (35 g polypepton, 20 g yeast extract, 5 g NaCl, 100 mg ampicillin and 10 mM IPTG per liter, pH7.5) and incubated for 3 h at 30°C. The cells were then collected by centrifugation and stored at -20°C until use.

Whenever necessary, the ts-scFv was purified as follows. The *E. coli* cells suspended in PBS were disrupted by a sonicator. The solution was centrifuged and the supernatant was applied for ammonium sulfate fractionation. Next, the ts-scFv fraction was further purified by ion-exchange chromatography using DEAE-TOYOPEARL and CM-TOYOPEARL columns.

2.3. Purification of large T antigen by ts-scFv immobilized column

Purified ts-scFv was immobilized to Sepharose using CNBr-activated Sepharose 4B (Pharmacia Biotech AB) according to a kit manual. Immobilized ts-scFv density was 2 mg-protein/ml-bed volume. The ts-scFv-Sepharose 4B (1 ml) was packed onto a column (ϕ 8 mm x 20 mm). The column was equilibrated with 10 mM potassium phosphate buffer (pH7.5) at 37°C, and then 2.5 ml of crude extract of 4×10^7 COS-1 cells was applied onto the column. After washing the column with 5 ml of the buffer, the binding proteins were eluted with the ice-cold buffer containing 0.5 M NaCl.

2.4. Analysis

The antigen-binding activity was measured by enzyme-linked immuno-sorbent assay (ELISA). The scFv samples were diluted with buffer, and then incubated for 1 h at the desired temperature. The diluted samples were applied for wells of an ELISA plate coated with the antigen, and allowed to stand for 1 h at the temperature. After the wells had been washed with tween-PBS solution, the bound scFv was detected with POD-conjugated anti-E tag antibody (Pharmacia Biotech AB). 2,2'-Azino-bis (3-ethylbenz-thiazoline-6-sulfonic acid) diammonium salt was used as a substrate for detection of POD activity, and the absorbance was measured at 405 nm.

3. RESULTS AND DISCUSSION

3.1. Design and screening of temperature-sensitive scFv

The designed helix-coil transition peptide, -(-Glu Ala Ala Ala Lys-)-, can reverse its conformation in response to temperature changes between 4 and 37°C. In particular, 80% of peptides take the helical form at 4°C and 60% of peptides take the random coil form at 37°C (n=3).[7] Thus, the conformation of the peptide becomes compact and the distance between the amino- and carboxyl-terminals of the peptide shortens at low temperatures. We concluded that the temperature-dependent conformational change in scFv had occurred in response to a change in linker peptide conformation and that this had led to a reduction in the antigen-binding activity (Fig. 1).

We first examined the effects of the helix repeat length on the antigen-binding activity and temperature responses. In order to do so, we prepared three different oligonucleotides corresponding to the peptides, which differed in terms of the helix repeat length (n=3-5, Fig. 2), and then incorporated them into the scFv gene (Fig. 3). When four or five helix repeat units (n=4,5) were used as the linker peptide, only a few *E. coli* clones expressed active scFv. However, these scFvs did not have a temperature-sensitive (ts) character. Thus, the length of these linker peptides may have been too long to fold as the active scFv. When three helix repeat units (n=3) were used, the clone number expressing active scFv increased to more than one hundred, and seven of these clones produced an scFv expressing apparent ts character. The DNA sequences of linker peptides of the ts-scFvs obtained were determined, and four clones showing different linker sequence were identified (Table 1). The ts-scFv from each clone had at least one sequence of the temperature-sensitive helix-coil transition peptide, although the number of amino acids residues varied in the range

Table 1. DNA sequence and amino acid sequence of each linker peptide

Sample	DNA sequence of linker peptide Amino acid sequence of linker peptide	Number of amino acid residue	Helix repeat unit	Antibody titer at 4°C*
#18	GGCGCC GAG GCA GCT GAG CCG GAA GCC GCA GCA AAG GCTAGC Glu Ala Ala Glu Pro Glu Ala Ala Ala Lys	14	1	1/4.8
#32	GGCGCC GAG GCA GCC GCC AAA GAG GCA GCA AAG GAA GCC GCA GCA AAG GCTAGC Glu Ala Ala Ala Lys Glu Ala Ala Lys Glu Ala Ala Ala Lys	18	2	1/8.3
#37	GGCGCC GAG GCA GCT GCA AAG GCTAGC Glu Ala Ala Ala Lys	9	1	1/16
#69	GGCGCC GAG GCA GCT GCC AAA GAG GGG CCG CTA AGG AAG CCG CAG CAA AGG CTA GCTAGC Glu Ala Ala Ala Lys Glu Gly Pro Leu Arg Lys Pro Gln Gln Arg Leu	20	1	1/36

*Antigen-binding activity of each scFv at 37°C was defined as 1.

of 9-20 residues. Antibody titers of the ts-scFvs at 4°C were measured by ELISA by comparing them with those at 37°C. The scFv from #69 clone (peptide sequence; Glu Ala Ala Ala Lys Glu Gly Pro Leu Arg Lys Pro Gln Gln Arg Leu) showed the maximum ts character, with the antigen-binding activity decreasing to 1/36 at 4°C when compared with that at 37°C. Thus, the ts-scFv from #69 clone was used for further study.

3.2. Characterization of ts-scFv

Figure 4 shows ELISA titration curves of usual scFv and ts-scFv (#69), when the incubation temperature for binding was 4 and 37°C. In the case of scFv prepared using a flexible linker peptide sequence of -(-Gly Gly Gly Gly Ser-)$_3$-, the slope of the titration curve became gentle at 4°C. This indicated that the affinity constant had slightly decreased at the low temperature. Nevertheless, scFv still possessed a strong affinity against the antigen at 4°C. On the other hand, ts-scFv adsorbed to the antigen very slightly at 4°C, whereas it exhibited strong affinity at 37°C. The binding activity of ts-scFv at 37°C was lower than that of usual scFv. Given that the slope of the titration curve was almost identical for scFv and ts-scFv, the affinity constant might not have changed between the scFvs at 37°C. The antibody titer of scFvs at 4 and 37°C was estimated using the titration curves, as shown in Fig. 4 (Table 1). In response to the temperature change, the binding activities of usual scFv and ts-scFv decreased to 1/2.4 and 1/36, respectively. The reduction in antibody titer for ts-scFv was 15-fold greater than usual scFv.

Figure 5 shows ELISA titration curves of ts-scFv at various temperatures. As can be seen, the binding activity gradually changed between 4 and 37°C. This may have coincided with the transition in helix-coil content.[7] The decreased binding activity at low temperatures was restored at 37°C.

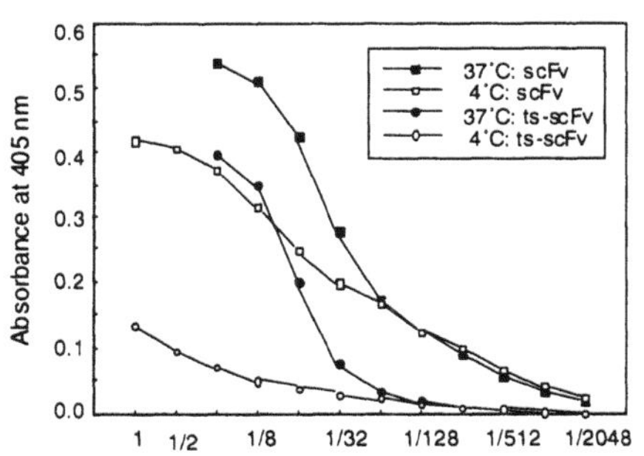

Fig. 4. ELISA titration curve of antigen binding activity of scFvs at 4 and 37°C

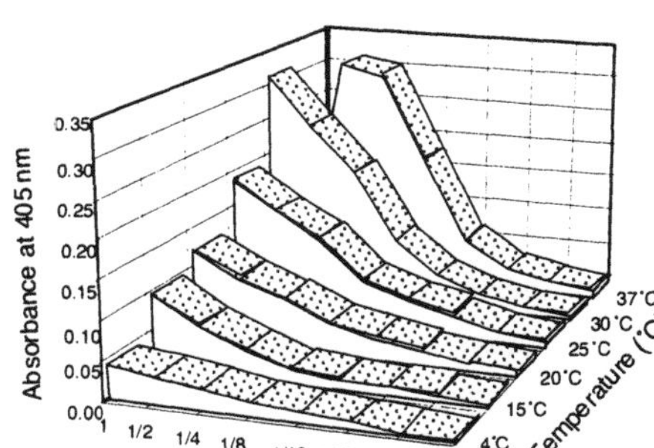

Fig. 5. ELISA titration curve of antigen binding activity of ts-scFv at various temperatures

The effect of ionic strength on antigen-binding activity was also examined. The binding activity of normal scFv was not affected by the addition of NaCl, whereas the ts-scFv exhibited an ionic strength dependency. The maximum binding activity was obtained without NaCl. The binding activity decreased as NaCl concentration increased. The binding activity in 0.5M NaCl was approximately 1/16 compared to that of without NaCl. The dependency of ionic strength on the antigen-binding activity was observed at both 4 and 37°C. The helical structure in the linker peptide may have been stabilized under the high ionic strength.

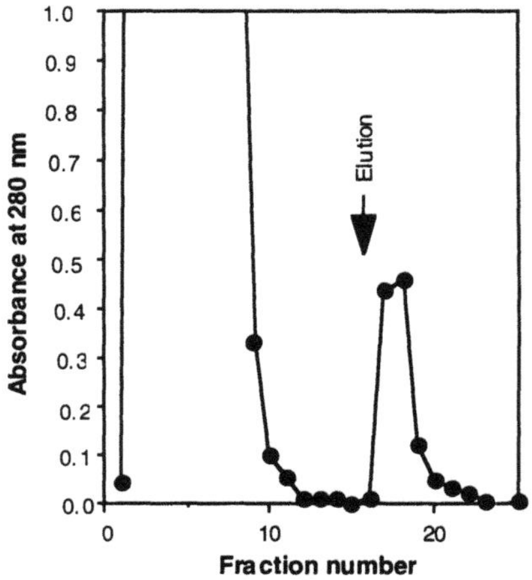

Fig. 6. Elution profile from ts-scFv Sepharose 4B column

3.3. Purification of antigen using ts-scFv column

Next, the ts-scFv from clone #69 was immobilized on Sepharose 4B to prepare an affinity column. The lysate of COS cells expressing large T antigen was applied onto the column at 37°C. After washing the column, ice-cold elution buffer containing 0.5M NaCl was added to elute the adsorbed proteins. Figure 6 shows the elution profile. About 350 μg of proteins corresponding to 1/80 of total protein loaded was eluted from the column by shifting to the elution condition. Judging from SDS-PAGE analysis for eluted samples, a major band of the eluted protein was stained at approx. 80kDa, and the protein band was identified as large T antigen by Western blotting. Although some minor bands were observed in the eluted sample, the specific activity was enhanced 70-fold.

REFERENCES

1. M. Kamihira, S. Iijima and T. Kobayashi: Bioseparation, 3 (1992) 185-188.
2. J. McCafferty, A.D. Griffiths, G. Winter and D.J. Chiswell: Nature, 348 (1990) 552-554.
3. J.D. Marks, A.D. Griffiths, M. Malmqvist, T.P. Clackson, J.M. Bye and G. Winter: Bio/technology, 10 (1992) 779-783.
4. G. Winter and C. Milstein: Nature, 349 (1991) 293-299.
5. T. Clackson, H.R. Hoogenboom, A.D. Griffiths and G. Winter: Nature, 352 (1991) 624-628.
6. G. Merutka and E. Stellwagen: Biochemistry, 29 (1990) 894-898.
7. V. Munoz and L. Serrano: J. Mol. Biol., 245 (1995) 297-308.

Bioseparation Engineering
I. Endo, T. Nagamune, S. Katoh and T. Yonemoto (Editors)

Stability enhancement of α-amylase by supercritical carbon dioxide pretreatment

Hwai-Shen Liu* and Yu-Chia Cheng

Department of Chemical Engineering, National Taiwan University, No.1, Sec. 4, Roosevelt RD., Taipei, 10617, Taiwan, R.O.C.
*Email: hsliu@ccms.ntu.edu.tw

Although 29% activity loss during powder pretreatment by CO_2 at 50℃, 2000 psi for 1 hour, α-amylase (EC 3.2.1.1) maintains 58% activity in water. That is about 41% activity kept based on the original activity before the treatment. Otherwise non-treated α-amylase almost loses its activity completely in a hour in water without buffer. This result provides a new method to improve enzyme stability.

1. INTRODUCTION

Enzyme stability is greatly influenced by the presence of water (Klibanov, 1989; Zaks and Russel, 1988). For example, the stability of α-amylase dissolved in water is very poor especially in low concentration without buffer. Many researchers focus on the methods that can promote the stability of enzymes. Among various possible methods of enzyme stability enhancement, immobilization (Brodelius, 1978; Sadhukhan *et al.*, 1990; Sadhukhan *et al.*, 1993), addition of various compounds (Asther and Meunier, 1990; Kalibanov, 1983; Violet and Meunier, 1989; Ward and Moo-Young, 1988; Windish and Mhatre, 1965) and chemical modification (Fretheim *et al.*, 1979; Tsuji, 1990) are frequently mentioned.

The enzyme α-amylase (1,4-α-D-glucanohydrolase, EC 3.2.1.1), widely used in the starch-to-fructose process, randomly hydrolyzes α-1,4 glucosidic linkages in polysaccharides into three or more α-1,4 linked D-glucose units to produce maltose or large oligo-saccharides (Boyce, 1986; Norman, 1981).

Supercritical fluids generally have similar density to liquids and similar viscosity to gases (Randolph, 1985). Thus, they are often recognized for their solvent power like liquids and diffusion capability in solids as gases. Among the supercritical fluids, supercritical carbon dioxide (abbreviated as SC-CO_2 hereafter) is the most frequently mentioned for its mild condition. That is, the critical pressure and temperature of 1070 psi and 31.3℃, repectively, make it suitable for various applications in food industry. Therefore, SC-CO_2 has attracted a great deal of attention for its use in the extraction of natural food substances (Taniguchi *et al.*, 1985 and 1987). In this report, a novel application of SC-CO_2 to enhance the stability of α-amylase is explored.

2. MATERIALS AND METHODS

2.1 Enzyme

Bacillus subtilis α-amylase powder (Merck, product 101329, EC 3.2.1.1) was used without further purification. None of protease activity was detected by the method Aderibigbe *et al.* (1990).

2.2 Enzyme Assay

One unit of α-amylase activity is defined as the amount of enzyme which can hydrolyze 1 mg starch in 25℃ at pH 6.8. Hydrochloride of 0.1 M is used as the stopping reagent.

Starch concentrations were determined by iodine test for various conditions. Iodine reacted with starch to give a dark blue complex that was measured with a spectrophotometer (Spectronic 20 Genesys) at a wavelength of 610nm. The absorbance of this starch-iodine complex (Hopkins and Bird, 1954) was checked for various iodine concentrations to ensure its linearity.

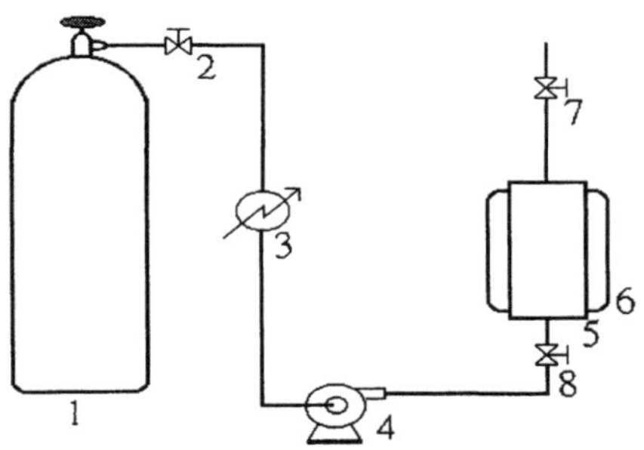

Figure 1. SC-CO_2 apparatus (1: CO_2 tank, 2: Valve, 3: Cooling, 4: Pump, 5: Vessel, 6: Heating jacket, 7: Output valve, 8: Input valve)

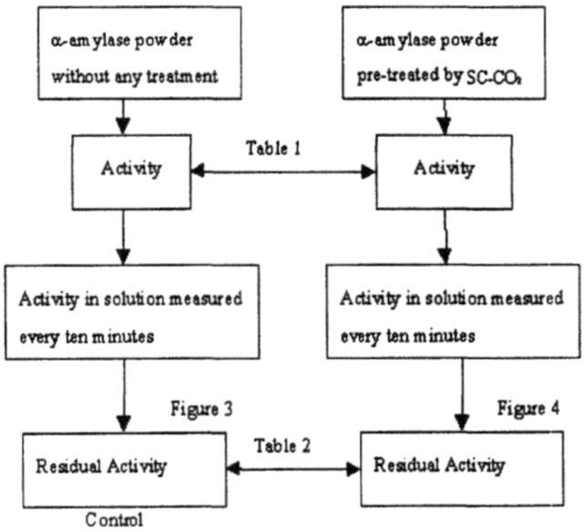

Figure 2. The flow sheet of experimental procedure

2.3 SC-CO_2 treatments

The experiments were carried out in a batch mode and the schematic setup is shown in Figure 1. The exposure time of α-amylase powder in the vessel under SC-CO_2 was set to one hour at various pressures and temperatures and then depressurized. The residual activity of the α-amylase (10 U/ ml) was measured at an atmospheric pressure. The activity of α-amylase solution was then measured every ten minutes to evaluate its stability. The entire procedure is diagrammed in Figure 2.

3. RESULTS AND DISCUSSION

3.1 Stability of α-amylase in solution

As shown in Figure 3, α-amylase was quite stable in phosphate buffer. However, without buffer, α-amylase lost more than 95% activity in one hour. *Bacillus subtis* α-amylase contains no disulfide linkages, due to the complete absence of cysteien and cystine (Fisher *et al.*, 1958). Therefore, *B. subtilis* α-amylase may easily lose its 3-D structure by the hydrogen linkages in the water. Further study on the effect of water on the structure of α-amylase and the interaction between active sites and water molecules may be necessary.

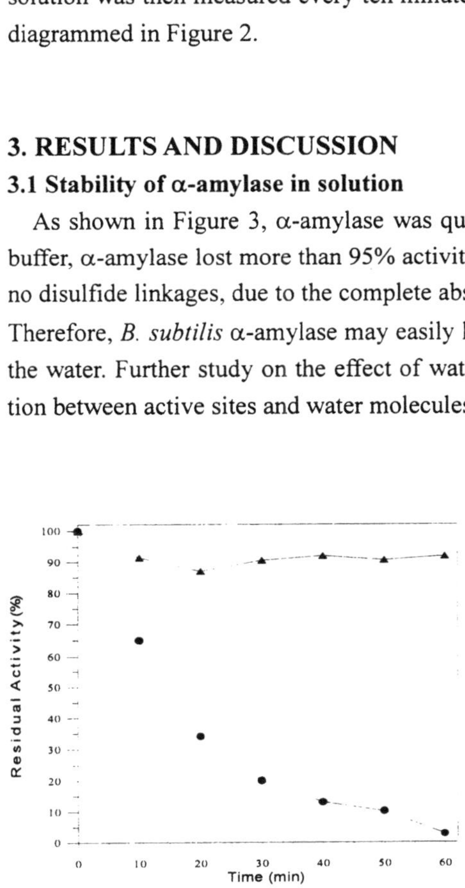

Figure 3 The stability of α-amylase solution (concentration: 10 U/ ml) under 25°C, 1 atm. (●: pH 6.8, without buffer ; ▲: pH 6.8, 0.05 M phosphate buffer)

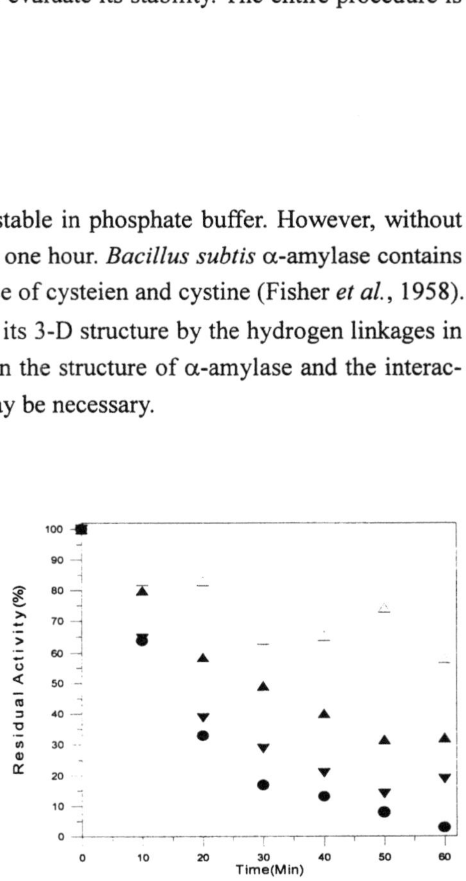

Figure 4 Stability of α-amylase treated by 50±0.5℃ SC-CO_2 for an hour in solution without buffer under 25°C, 1 atm. (●: without treatment ; △: 2000psi; ▲: 2500psi; ▼: 4000psi)

Table 1
The effect of $SC\text{-}CO_2$ treatment on α-amylase at 50℃ and various pressures

Pressure (psi)	Activity immediately after treatment (%)	Residual activity* in solution after one hour (%)
1500	77.36	32.13
2000	70.85	58.08
2500	80.78	44.03
3000	84.97	17.58
3500	61.12	19.29
4000	69.47	18.81

*based on the activity immediately after treatment

3.2 Effect of $SC\text{-}CO_2$ treatment on the activity of α-amylase

The residual activities of α-amylase pre-treated by $SC\text{-}CO_2$ at 50±0.5℃ and various pressures ranged from 60 to 80% as shown in Table 1. The activity loss may be due to the active site deformation caused by the high pressure and/or depressurization. Some research also showed that the residual activity may be higher than 80% if operated at lower temperature or less water content (Taniguchi et al., 1987; Yang and Yang, 1997).

3.3 Effect of $SC\text{-}CO_2$ treatment on the stability of α-amylase solution

When pre-treated by $SC\text{-}CO_2$ at 50±0.5℃ and 2000 psi, α-amylase solution without buffer can lower its activity loss to 40% , shown in Figure 4, based on the activity immediately after treatment. This is a significant improvement in stability compared with more than 95% loss for non-pretreated α-amylase in Figure 3. We hypothesize that the suitable pressure and temperature induce the stability enhancement. Perhaps high pressure makes the structure of α-amylase compact and rigid. Hence, it becomes more stable. However, when the higher pressure is applied (3000-4000 psi), it may be too harsh for the enzyme. That's why the stability didn't improve further with higher pressure. As for temperature, it may provide enzyme molecules kinetic energy to prevent over-deformed. That's why α-amylase pre-treated by 40℃ $SC\text{-}CO_2$ didn't enhance the stability (Result not shown).

4. CONCLUSION

In Table 2, a comparison of α-amylase with proper pretreatment and the original one is summarized. Although the enzyme powder lost about 30% activity during the pretreatment,

Table 2
Comparison of α-amylase stability between pretreated enzyme and original one

	Activity(%)	
	fresh solution	Dissolved in water after one hour
α-amylase without any treatment	100	2.91
α-amylase pretreated by 50°C, 2000 psi SC-CO_2	70.85	41.15

α-amylase gained significant improvement in stability as it dissolves in water. That is, pretreated α-amylase solution maintained 41 % activity while original one lost 97% activity in an hour. That turns out to be about 14 times increase in stability as listed in Table 2. Some may suspect the possibility of protease destroyed during the pretreatment so that the stability of α-amylase was obtained. However, no protease activity was found as mentioned in the section of Materials and Methods. Even if it is true, this pretreatment also presents a way to protect enzyme activity. Anyway, though the detail mechanism involved in this method of stability enhancement is not clear, it does provide a novel method to improve enzyme stability.

5. ACKNOWLEDGEMENTS

The financial support of National Science Council is gratefully acknowledged.

REFERENCES

1. Aderibigbe, E. Y., Schink B., and Odunfa, S. A. , (1990). *Food Microbiol.,* **7**, 281-293
2. Asther, M. and Meunier, J. C. (1990). *Enzyme Microb. Technol.*, **12**, 902-905
3. Boyce, C. O. L. (1986). *Novo's handbook of practical biotechnology*, p. 35, Novo industri A/S
4. Brodelius, P. (1978). *Adv. Biochem. Eng.*, **10**, 75-129
5. Fisher, E. H., Summerwell, W. N., Junge, J. M. and Stein, E. A. (1958). *Proceedings of Symposium VIII, IVth International Congress of Biochemistry, Vienna.* Pergamon Press.
6. Fretheim, K., Iwai, S. and Feeny, R. E. (1979). *Int. J. Peptide. Potein. Res.*, **14**, 451-454
7. Hopkins, R. H. and Bird, R. (1954). *Biochem. J.*, **56**, 86-89

8. Klibanov, A. M. (1983). *Adv. Appl. Microbiol.*, **29**, 1-28
9. Klibanov, A. M. (1989). *TIBS*, **14**, 141-144
10. Norman, B. E. (1981). New developments in starch syrup technology. In: *Enzymes and processing*, Birch, G. G., Blakebrough, N., and Parker, K. J., eds pp15-50, England: Applied Science Publishers Ltd.
11. Randolph, T. W., Blanch, H. W., Prausnitz, J. M. and Wilke, C. R. (1985). *Biotechnol. Lett.*, **7**, 325-328
12. Sadhukhan, R. K., Manna, S., Roy, S. K. and Chakrabarty, S. L. (1990). *Appl. Microbiol. Biotechnol.*, **33**, 692-696
13. Sadhukhan, R., Roy, S. K. and Chakrabarty, S. L. (1993). *Enzyme Microb. Technol.*, **15**, 801-804
14. Taniguchi, M., Kamihira, M. and Kobayashi, T. (1987). *Agric. Biol. Chem.*, **51**, 593-594
15. Taniguchi, M., Nomura, R., Kijima, I. and Kobayashi, T. (1987). *Agric. Biol. Chem.*, **51**, 413- 417
16. Taniguchi, M., Tsuji, T., Shibata, M. and Kobayashi, T. (1985). *Agric. Biol. Chem.*, **49**, 2367-2372
17. Tsuji, R. F. (1990). *Biotechnol. Bioeng.*, **36**, 1002-1005
18. Violet, M. and Meunier, J. C. (1989). *Biochem. J.*, **263**, 665-670
19. Ward, O. P. and Moo-Young, M. (1988). *Biotech. Adv.*, **6**, 39-69
20. Windish, W. W. and Mhatre, N. S. (1965). *Adv. Appl. Microbiol.*, **7**, 273-304
21. Yang, J. C. and Yang, X. M. (1997). *The 4th International Symposium on Supercritical Fluids*, pp 139-141, Sendai, Japan
22. Zaks, A. and Russel, A. J. (1988). *J. Biotechnol*, **8**, 259-270

Bioseparation Engineering
I. Endo, T. Nagamune, S. Katoh and T. Yonemoto (Editors)

Behavior of Monodispersed Oil-in-Water Microsphere Formation Using Microchannel Emulsification Technique

Jihong Tong, Mitsutoshi Nakajima*, Hiroshi Nabetani and Yoji Kikuchi
National Food Research Institute, Ministry of Agriculture, Forestry and Fisheries, Kannondai 2-1-2, Tsukuba, Ibaraki 305-8642, Japan

Super-monodispersed oil-in-water (O/W) microspheres (MS) were produced by microchannel (MC) emulsification technique. To investigate the behavior of the O/W-MS formation, three kinds of surfactant were used in the MC emulsification process. An MC plate with 8.9 μm in equivalent diameter was employed. It was found that the nonionic and anionic surfactants could be used for the super-monodispersed O/W-MS production. The average droplet diameter was about 30 μm with a standard deviation less than 1 μm. The behavior of the O/W-MS formation, the MS size and its distribution were discussed.

Keywords:
Microsphere, Monodispersed MS, Microchannel emulsification, Surfactant.

1. INTRODUCTION

Microspheres (MS), which are emulsion cells or solid particles dispersed in a continuous phase, have been utilized in various industries such as foods, cosmetics and pharmaceuticals, etc. Using the conventional methods of emulsion production, the emulsions (or MS) produced are usually considerably polydispersed over a wide range. Recently Kawakatsu *et al.* (1) have proposed a novel microchannel (MC) emulsification technique for super-monodispersed MS production. Using the MC emulsification technique, both oil-in-water (O/W) and water-in-oil (W/O) MS with monodispersibility have been produced and the characterization of the produced MS has also been investigated (2, 3).

It is considered that a surfactant plays a very important role in an emulsification process. Surfactant lowers the interfacial tension and facilitates emulsion formation. Surfactant is supposed to induce repulsive force between droplets and to stabilize the emulsion (5, 6). In case of this study, surfactant is considered to be affecting the

* Corresponding author

hydrophobic property of the MC surface, which is directly influencing the behavior of the MS formation. Therefore, we investigated the behavior of the O/W-MS formation, MS size and its distribution by using different kinds of surfactants.

2. MATERIALS AND METHODS

2.1. Reagents

High-oleic sunflower oil (triolein, >90% purity) was obtained from Nippon Lever B.V., Tokyo, Japan. Sodium oleate and polyoxyethylene (20) sorbitan monooleate (Tween 80, HLB: 15.0) were purchased from Wako Pure Chemical Ind., Osaka, Japan. Di-2-ethylhexyl sodium sulfosuccinate (AOT) was purchased from Sigma Chemical Co., St. Louis, MO, USA. All materials were reagent grade and were used without further purification.

2.2. Apparatus and procedure

The silicon MC plate with partition walls between the channels for both sides of the terrace is shown in Fig. 1 (a). Its dimension is 15 mm x 15 mm x 0.5 mm. 600 channels around the 4 side with 8.9 μm in equivalent diameter were formed on the silicon plate. The

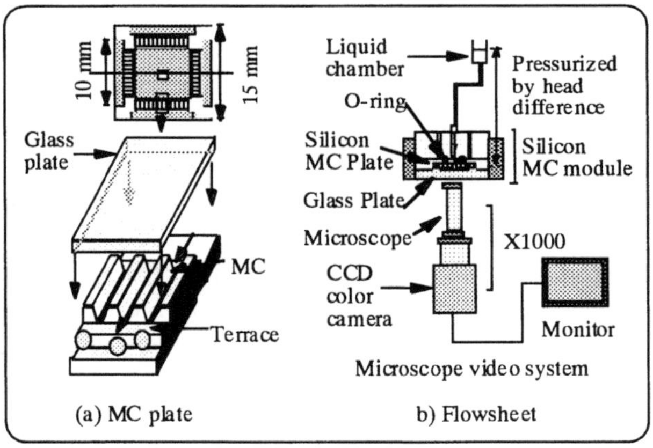

Fig. 1 Experimental apparatus of the MC emulsification technique, a) MC plate; b) Flowsheet.

flowsheet of the experimental apparatus is given in Fig. 1 (b). A module installed with an MC plate adhering to a flat glass plate was filled with a water phase. An oil phase chamber contacting to the module by a silicone tube offered the dispersed phase to the module. A microscope video system and a monitor were employed to record and observe the MC emulsification process. The oil phase was pressed into the module by lifting the oil phase chamber. When the head difference between the chamber and the module was large enough, the oil phase broke through the MC and began to form MS. The pressure applied at this point was defined as breakthrough pressure. The behavior of the MS formation was analyzed from the video images recorded by a 3CCD video camera with about 1000x enlargement, while the MS size and its distribution were determined by counting over 200 droplets by using a Macintosh computer. The interfacial tension was measured by an automatic interfacial tensiometer (PD-W, Kyowa Interface Science Co., Saitama, Japan) with pendant drop method. All experimental runs were carried out at room temperature.

3. RESULTS AND DISCUSSION

Two anionic surfactants were used in this study. AOT was dissolved into the oil phase, while sodium oleate was used by dissolving it into the water phase. For each surfactant, several concentration conditions within 0.05 - 1.0 wt.% were tested, and the breakthrough pressure for each condition was recorded. The effect of the concentration of AOT and sodium oleate on the interfacial tension and the breakthrough pressure is shown in Fig. 2.

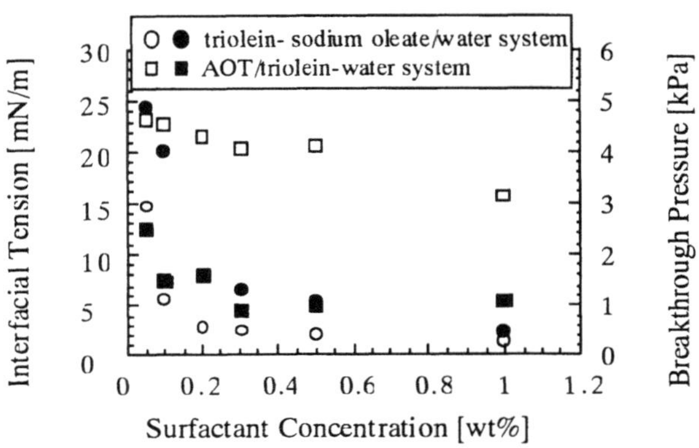

Fig. 2 Effect of surfactant concentration on interfacial tension and breakthrough pressure, open keys: interfacial tension, solid keys: breakthrough pressure.

It was found that the interfacial tension and the breakthrough pressure decreased with the increase in surfactant concentration. It is considered that the breakthrough pressure strongly depends on the interfacial tension. At the surfactant concentration higher than c.m.c., the interfacial tension and the breakthrough pressure became two constant. Although the interfacial tension data of two systems shows remarkable difference, the breakthrough pressure did not show so large difference especially at higher concentration over 0.2 wt.%.

For the same experimental systems, the effect of surfactant concentration on the MS average diameter and the standard deviation is shown in Fig. 3. The average droplet diameter was found to be changed slightly within the concentration range investigated. It tends to give smaller MS diameter at lower surfactant concentration. This phenomenon seems to be conflictive, since higher interfacial tension occurred in lower surfactant concentration, thereby larger droplet would be created generally. However, the experimental runs gave the result of smaller droplet actually. It is supposed that this reason may be related to the inflation process of the MS outside the channel and the detachment mechanism of the formed MS from the terrace.

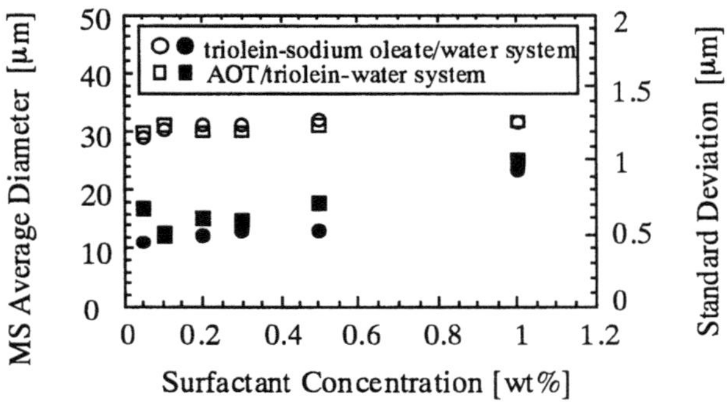

Fig. 3 Effect of surfactant concentration on MS average diameter and standard deviation open keys: MS average diameter, solid keys: standard deviation.

Within the concentration range studied, it is found that AOT containing MS had larger standard deviation than the MS produced by using sodium oleate. From the molecular structures of AOT and sodium oleate, sodium oleate has a C18:1 chain with an unsaturated bond, while AOT has two shortest main chains with two sub-chains, it means that the cross section of the hydrophobic tails of AOT is probably larger than that of its hydrophilic group. AOT has been used to form reversed micelles easily, a kind of W/O microemulsion used for protein extraction (4). AOT could not function as well as sodium oleate did in this study, probably due to the differences of the molecular structure, the hydrophobic property and the interfacial tension.

On the other hand, polyoxyethylene (20) sorbitan monooleate (Tween 80, HLB: 15.0) was used as a nonionic surfactant in this study. When it was dissolved into the oil phase at 0.3 wt.%, good behavior of the O/W-MS formation was obtained. The average droplet diameter of the produced MS was 31.8 μm and the standard deviations were 1.02 μm. Tween 80 was also dissolved into both the oil and water phases at the concentration of 0.3 wt.%, so that the mass transfer of Tween 80 between two phases was reduced during the emulsification process. In this case, the O/W-MS production was also performed well and the average droplet diameter was 29.8 μm and the standard deviation was 0.42 μm. Comparing to the data obtained by dissolving into only the oil phase, the diameter was a little smaller and the monodispersibility was a little better. This is probably attributed to the higher diffusion rate and adsorption level of the surfactant to the formed MS interface when Tween 80 was also dissolved in the water phase.

Fig. 4 shows the drop size distribution for the different surfactants in the same concentration condition. It apparently shows the monodispersibility of the O/W-MS formed with the anionic and nonionic surfactants by using the MC emulsification technique.

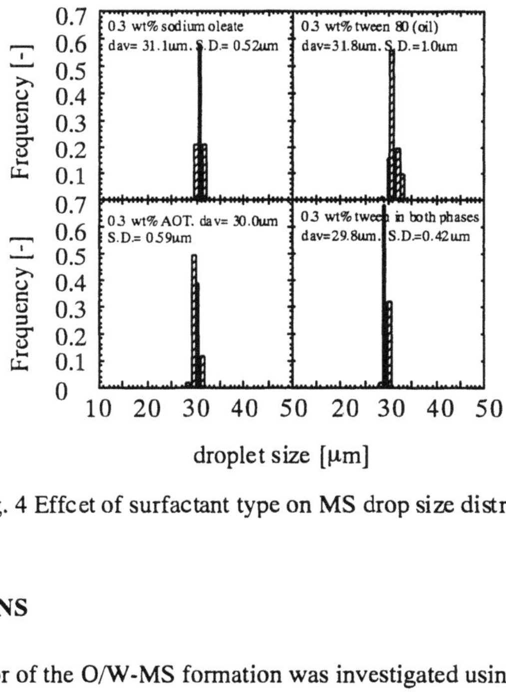

Fig. 4 Effcet of surfactant type on MS drop size distribution

4. CONCLUSIONS

The behavior of the O/W-MS formation was investigated using 3 kinds of surfactant in the MC emulsification process by employing an MC plate with 8.9 μm in equivalent diameter.

It was found that the interfacial tension affects the breakthrough pressure when MS began to form in the MC emulsification process.

When the anionic and nonionic surfactants were used, the monodispersed O/W-MS production was succeeded and the average droplet diameter was about 30 μm with a standard deviation less than 1 μm. It showed the monodispersibility of the produced O/W-MS.

Acknowledgment: This work was supported by Program for Promotion of Basic Research

Activities for Innovative Biosciences of Japan (MS-Project).

REFERENCES

1. Kawakatsu, T., Kikuchi, Y. and Nakajima, M. *J. Am. Oil Chem. Soc.*, 74, 317-321(1997).
2. Kawakatsu, T., Komori, H. Oda, N. and Yonemoto, T. *Kagakukogaku Ronbunshu*, 24, 313-317 (1998).
3. Kobayashi, I., M, Nakajima, J. Tong, T. Kawakatsu, H. Nabetani, Y. Kikuchi, A. Shohno and K. Satoh, accepted, *Food Sci. Technol. Res.*
4. Tong, J. and S. Furusaki, *Sep. Sci. Tech.*, 33, 899-907 (1998)
5. Schubert, H. and H. Armbruster, *Intel. Chem. Eng.*, 32, 14-28 (1992)
6. Walstra, P., Dispersed systems: basic consideration, Food Chemistry, 3rd Ed., edited by Owen R. Fennema, Marcel Dekker, Inc., 95-155 (1996)

Chapter 4

Bioseparation Engineering for Global Environment

Bioseparation Engineering
I. Endo, T. Nagamune, S. Katoh and T. Yonemoto (Editors)

Domestic wastewater treatment using a submerged membrane bioreactor

Xia Huang, Ping Gui and Yi Qian

Environment Simulation and Pollution Control State Key Joint Laboratory
Department of Environmental Science and Engineering
Tsinghua University, Beijing 100084, China

In the present study, performance of a biological reactor submerged with a hollow fiber membrane module was investigated for treating domestic wastewater. Five runs with hydraulic retention time (HRT) of 5 h, sludge retention times (SRTs) of 5, 10, 20, 40 and 80 d respectively, were conducted. The submerged membrane bioreactor process was capable of achieving over 90% removals both for COD and NH_3-N on the average almost independent of SRT. The maximum COD and NH_3-N loadings obtained in the study were 4.0 kg-COD m^{-3} d^{-1} and 0.18 kg-NH_3-N m^{-3} d^{-1}, respectively. Sludge concentration in the bioreactor increased with prolonged SRT. Sludge yield coefficient and endogenous coefficient was calculated as 0.25 kg-VSS kg-COD^{-1} and 0.04 d^{-1}, being similar to that of the conventional activated sludge process.

Key Words: Submerged membrane bioreactor, Domestic wastewater treatment, COD removal, NH_3-N removal, Sludge retention time, Sludge concentration

1. INTRODUCTION

With the progress of membrane technology, application of membrane separation in wastewater treatment has received high attention in recent decades. Membrane bioreactor is a combination process of biological reactor with membrane separation. In the combination process, due to efficient separation performance of membrane, biosolids with high concentration can be retained within bioreactor, which enables operation in high organic loading and makes the equipment compact. Moreover, high quality effluent can be also obtained. For above advantages, membrane bioreactor is commonly considered as an innovative technology for wastewater treatment and reclamation. Study on application of membrane bioreactor in treating domestic wastewater, night soil wastewater, and industrial wastewater has attracted a great attention [1-3].

Several types of membrane bioreactor have been investigated. The conventional type is that a membrane module is allocated outside a bioreactor and a circulation pump is used to generate cross flow over the membrane surface [4]. This type of membrane bioreactor is simple and easy to be operated. However, quite amount of energy is consumed to generate a high circulation velocity over membrane surface to maintain a high filtration flux. To eliminate the disadvantages of the conventional type, a new type of membrane bioreactor was proposed [5]. In this process, a membrane module is submerged in a bioreactor, and the effluent is extracted by a suction pump so that it is more compact and extra energy is not required.

The purpose of the present study was to investigate the performance of this submerged membrane bioreactor for domestic wastewater treatment at different sludge retention times (SRTs). Sludge growing and kinetic parameters have been studied as well.

2. EXPERIMENTAL

2.1. Experimental system and conditions

A schematic diagram of the experimental system consisting of an activated sludge bioreactor, in which a membrane module is submerged, is shown Figure 1. Activated sludge bioreactor is a rectangular tank of 900 mm in length, 120 mm in width and 1100 mm in height, separated into two parts by a plate. The membrane module used in the study is a plate of hollow fiber membranes made of polyethylene with the pore size of 0.1 μm and the total filtration area of 4 m^2, which can be operated at a transmembrane pressure lower than atmospheric pressure. Air aeration supplied from aeration pipes underneath the membrane module generates cross flow along the membrane surface by an air lift effect to hinder deposition of suspended solids on membrane surface.

Domestic wastewater taken from Tsinghua campus was used in the study. After passing through a fine screen to remove rough suspended solids, domestic wastewater flowed into a wastewater storage tank and then was pumped up to activated sludge bioreactor. Membrane effluent was intermittently extracted by a suction pump. The suction time and cease time was 13 min and 2 min.

The trial was conducted in five runs with different SRTs. The experimental conditions for each run are shown in Table 1. Hydraulic retention time (HRT) of bioreactor was constantly kept at 5h on the whole experimental period.

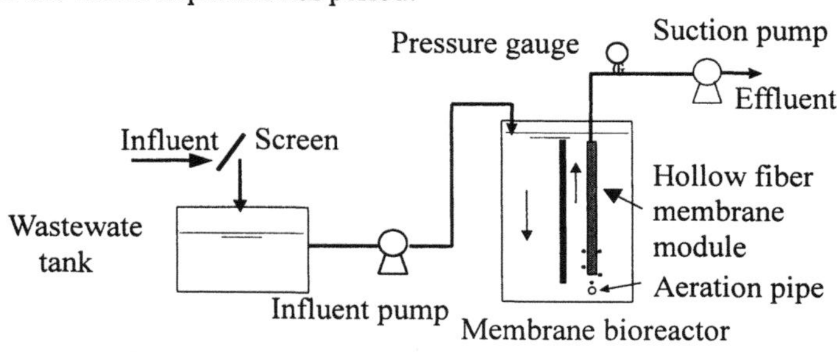

Figure 1. Schematic diagram of experimental apparatus.

Table 1
Experimental conditions

Items	Run-1	Run-2	Run-3	Run-4	Run-5
SRT (d)	5	10	20	40	80
HRT (h)			5		
Operation time (d)	70	43	120	140	45
DO in bioreactor (mg l^{-1})			4 – 5		
Influent temperature (℃)	9 - 17	16 – 21	19 – 21	19- 21	9 - 17
Permeating flux ($l\ m^{-2}\ h^{-1}$)			5		
Suction time/cease time			13 min/2 min		

2.2. Analytical items and methods

The analytical methods from Chinese NEPA Standard Methods were adopted for measurements of chemical oxygen demand (COD), ammonia nitrogen (NH_3-N) and pH in the influent, bioreactor effluent and membrane effluent, total suspended solids (SS) and volatile suspended solids (VSS) in bioreactor, respectively.

3. RESULTS AND DISCUSSION

3.1. COD removal performance

Figure 2 shows variations of COD concentrations of the influent, bioreactor effluent and membrane effluent during the five runs. The COD concentration of the bioreactor effluent was measured by detecting the supernatant of the mixed liquor after centrifuged at 4000 rpm and 15 min. On the whole experimental period, over 400 days, regardless of the wide fluctuation of influent COD from 40 to 800 mg l^{-1} and change of SRT, all of the membrane effluent COD were lower than 20 mg l^{-1}, which could meet the water quality standard for reuse issued by the Ministry of Construction of China. Membrane separation played an important role in keeping low and stable effluent COD.

On the other hand, bioreactor effluent COD varied from 10 - 100 mg l^{-1} with change of SRT. Along with prolonged SRT from 5 d to 20 d, bioreactor effluent COD decreased firstly and then increased if SRT was further prolonged. Higher COD concentrations of bioreactor effluent appearing at conditions of shorter SRT and longer SRT, respectively, might be attributable to incomplete decomposition of organic components in raw wastewater and accumulation of large molecular metabolites. Similar results concerning accumulation of metabolites at longer SRT have been also reported by several authors with different conditions [5,6].

Difference of about 20 - 30 mg l^{-1} between bioreactor effluent COD and membrane effluent

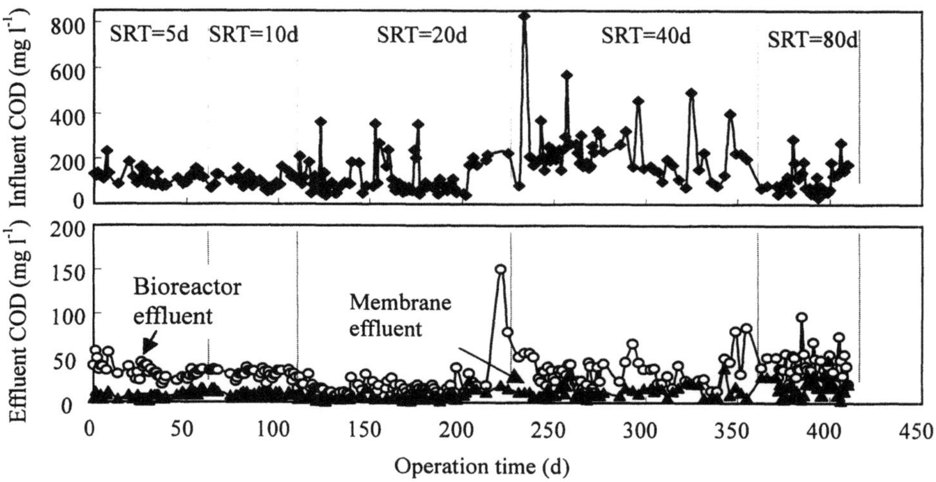

Figure 2. Variations of influent COD and effluent COD concentrations with different SRTs.

COD indicates that membrane could expel a fraction of dissolved COD components with relatively large molecular weight.

On the average, the COD removal efficiencies both for the total process and the bioreactor were over 90% and 75%, respectively.

As described above, since the influent COD concentrations varied largely, the influent volumetric COD loading changed along with even at the same HRT condition. The influence of COD loading on COD removal rate is shown in Figure 3. A linear relationship was confirmed. This result implied that COD components flowing in bioreactor could be effectively removed in membrane bioreactor process, even the volumetric COD loading was up to 4 kg-COD m^{-3} d^{-1}. For the conventional activated sludge process, COD loading is in the range of 0.6 to 1.2 kg-COD m^{-3} d^{-1}. The maximum COD loading obtained in the study was 3 to 6 times that.

The intercept of the line on the abscissa was about 0.13 kg-COD m^{-3} d^{-1}, which represents the part of poor biodegradable organic components in the influent.

3.2. NH_3-N removal performance

As shown in Figure 4, excellent NH_3-N removal performance could be also achieved.

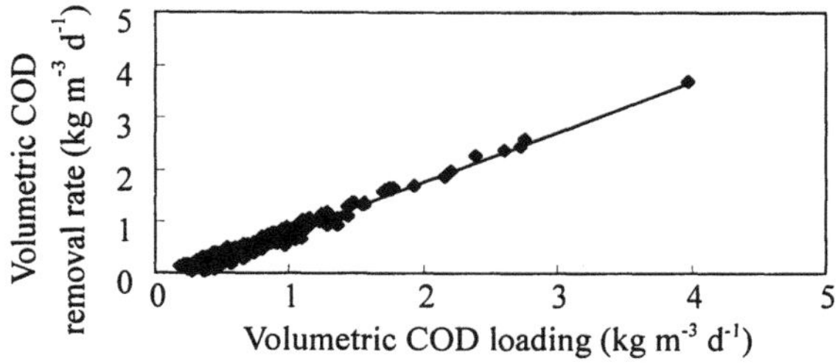

Figure 3. Influence of COD loading on COD removal rate.

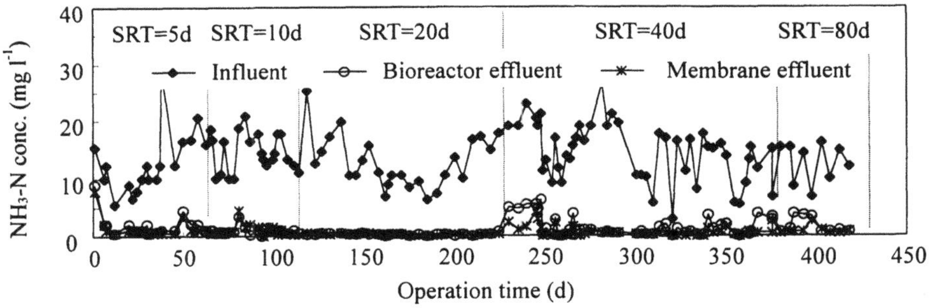

Figure 4. Variations of influent and effluent NH_3-N concentrations with different SRTs.

Effluent NH_3-N concentrations of bioreactor and membrane were quite low, being in 0 - 8.8 mg l^{-1} and 0 - 7.7 mg l^{-1}, respectively, even if influent NH_3-N concentrations varied largely from 3 mg l^{-1} to 29 mg l^{-1}. No obvious difference in NH_3-N concentrations between bioreactor effluent and membrane effluent was observed, which implied that NH_3-N removal is mostly contributed by biological reaction in bioreactor and NH_3-N molecular is too small to be cut off by membrane.

On the average, the NH_3-N removal was over 90% regardless of change of SRT.

A linear relationship between the volumetric NH_3-N removal rate and NH_3-N loading was also observed as shown in Figure 5. It suggested that NH_3-N in the influent could be removed well even the NH_3-N loading was up to 0.18 kg-NH_3-N m^{-3} d^{-1}.

3.3. Sludge growing and kinetic parameters

Sludge amount retrained in bioreactor is an important factor affecting the treatment capacity of a biological system. Mean SS and VSS concentrations, when the steady state was reached at different SRTs, are shown in Table 2. It was unambiguous that with increased SRT, sludge concentration increased. Sufficient sludge concentration will ensure good performance in COD removal and better effluent quality.

Based on the Lawrence-McCarty modeling [7], equation (1) can be used to describe the relationship between the mean sludge retention time θ_c of bioreactor and the organic removal rate $-r_s$ (kg-COD kg-VSS^{-1} d^{-1}) when membrane bioreactor process reaches the steady state at each SRT condition.

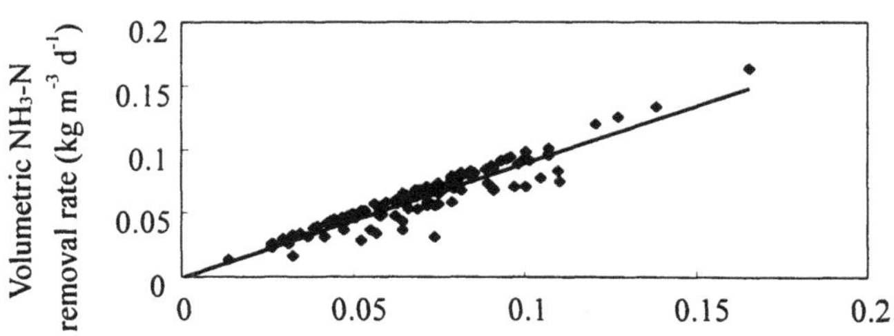

Figure 5. Influence of NH_3-N loading on NH_3-N removal rate.

Table 2
Mean sludge concentrations in bioreactor at different SRTs.

SRT (d)	5	10	20	40
SS ($g\ l^{-1}$)	1.5	2.3	3.0	7.0
VSS ($g\ l^{-1}$)	0.9	1.2	1.7	4.7
VSS/SS (-)	0.6	0.52	0.57	0.67

$$1/\theta_c = -Y r_s - b \qquad (1)$$

Where Y is true sludge yield coefficient and b is endogenous decay coefficient.

Through interrelation calculation using experimental data obtained in the study, a linear relationship between the reciprocal of the mean sludge retention time ($1/\theta_c$) and organic removal rate ($-r_s$) was confirmed. Consequently, Y and b was estimated as 0.25 kg-VSS kg-COD^{-1} and 0.04 d^{-1}. For the conventional activated sludge process, Y and b is usually in the ranges of 0.25 - 0.4 kg-VSS kg-COD^{-1} and 0.04 - 0.075 d^{-1}, respectively [7]. It was proven that kinetic parameters for sludge in membrane bioreactor approached the lower limit of the values for the conventional activated sludge process.

4. CONCLUSIONS

Excellent COD and NH_3-N removal performances can be obtained by using membrane bioreactor for treating domestic wastewater. The removals both for COD and NH_3-N were over 90% on the average regardless of the wide variation of influent COD and change of SRT.

The maximum COD and NH_3-N loadings obtained in the study were 4.0 kg-COD m^{-3} d^{-1} and 0.18 kg-NH_3-N m^{-3} d^{-1}, respectively.

Sludge concentration in the bioreactor increased with prolonged SRT. Kinetic analysis showed that the sludge yield coefficient and endogenous coefficient was 0.25 kg-VSS kg-COD^{-1} and 0.04 d^{-1}, being similar to that of the conventional activated sludge process.

REFERENCES

1. K. Brindle and T. Stephenson, Biotecn. Bioeng., 49(1996)601.
2. M.D. Knoblock, P.M. Sutton, P.N. Mishra, K. Gupta and A. Janson, Water Environment Research, 66(1994)133.
3. E. Trouve, V. Urbain and J. Manem, Wat. Sci. Tech., 30(1994)151.
4. Y. Magara and M. Itoh, Wat. Sci. Tech., 23(1991)1583.
5. K. Yamamoto, M. Hiasa, T. Mahmood and T. Matuso, Wat. Sci. Tech., 21(1989)43.
6. S. Chaize and A. Huyard, Wat. Sci. Tech., 23(1991)1591.
7. X.Sh.Gu, Mathematical modeling for biological wastewater treatment (in Chinese), Tsinghua University publishing house, Beijing, China, 1993.

Bioseparation Engineering
I. Endo, T. Nagamune, S. Katoh and T. Yonemoto (Editors)

Biosorption of Heavy Metal Ion with Penicillin Biomass

Tan Tianwei Chen Peng

(Department of Biochemical Engineering, Beijing University of Chemical Technology, Beijing 100029, P.R. China)

The adsorption of heavy metal ions with waster biomass from penicillin industry was studied in this paper. The parameters such as pH , concentration of salt and size of particle were investigated. When the pH is higher than 5.0, which is near to the pKa of chitosan , the high adsorption capacity could be obtained. Low salt concentration (<5%) would improve the adsorption of heavy metal ions. The comparison of the penicillin mycelium adsorbent with CM—Sephadex C-25 and chelated resin D-751 was made. The adsorption isotherms could be described by Freundhlish and Langmuir models. The desorption could be achieved by 0.5 M HNO3 and the biomass adsorbent could be used many times.

1. INTRODUCTION

Many industries especially plating, battery, pigment and so on produce a large amount of waster water which contains many heavy metal ions such as Cu^{2+}, Ni^{2+} and Cr^{3+} and these heavy metal ions often cause serious toxicity to human and other form of life(1). The removal of such heavy metal ions could be achieved by different methods ,for example, precipitation ,ion exchange and so on(2). These methods have some disadvantages , for example, precipitation often produces many precipitates which are difficult to treat, traditional ion exchange resins have low selectivity and the desorption is difficult .
Fermentation industries such as citric and penicillin industries produce large amount of waste biomass.Now the waste biomass are used as animal feed or are discarded, which often cause pollution problem. The use of microorganism in heavy metal ions removal and recovery process has been documented(3), which has high selectivity, low cost and easily regeneration. Mycelical waster of microbial origin from fermentation industry has many compositions such as chitosan and glucan , in which some groups such as NH2 or COOH have high adsorption capacity for heavy metal ion (4) . In this paper the adsorption of heavy metal ions with penicillin mycelium which contains large amount of chitosan was studied. Some other ion exchange resins are used as comparison.

2 MATERIALS AND METHODS

2.1 Chemicals

CM-Sephardex C-25 and Prototype Gel are kind gifts from Dr. Cecilia Pang of Pharmacia Biotech. D-751 chelated resin was bought from Shanghai Resin Plant(China). CuSO4 and NiSO4 and ZnSO4 are analytic grade.

2.2Biomass pretreatment

Waste biomass of penicillin chrysoganum was obtained from North China Pharmacitic Co. The pretreatment is as following: 1 kg biomass(dry) was added into 10 L 0.5 M NaOH and stirred at 30 ℃ for 3 h.The suspension was filtrated and the biomass was washed with water to pH7.0 and was dried under vacuum and was blended. The dry biomass was sizing by a mesh before use.

2.3 Metal ion analyses

The metal ion Cu^{2+} was analyzed with adsorption spectrophotometer according to method(5). The Ni^{2+} and Zn^{2+} was determined by method(6)

2.4 Biosorption experiment

25ml heavy metal ion solution (metal ion concentration 100 mg/ L) was added in a 50 ml flask. The pH was regulated to pH4.0--5.0 by 1M NaOH or 1M HCl. 0.1g dry biomass was added and the mixture was stirred at 30°C for 12 h. The metal ion concentration in the supernatant was determined and the adsorption capacity was calculated by mass balance. Q presents the up take of metal ion (mg/g dry biomass).

3. RESULTS AND DISCUSSION

3.1 Sorption isotherms

The sorption isotherms were shown in Fig.1 For the Ni^{2+}, 30 minutes will be taken to reach equilibrium, while for Cu^{2+} and Zn^{2+}, it took 60 minutes and 120 minutes, respectively. The time for biomass sorption is langer than that in normal ion exchange. Maybe the diffusion resistance of metal ions in biomass is higher than that in high pore ion exchange resin.

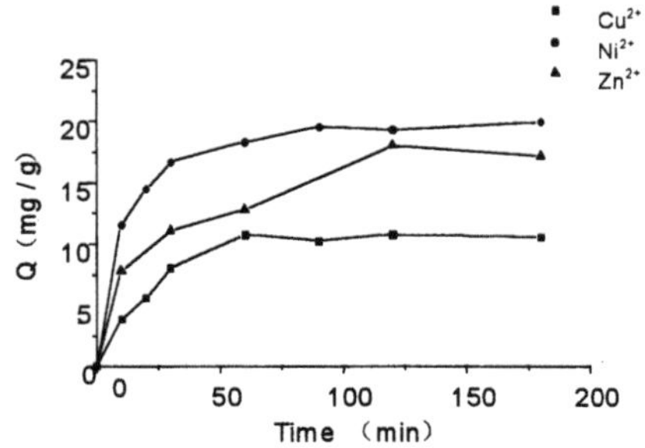

Fig. 1 Sorption isotherms of metal ions

pH:Cu^{2+} 4.0,Ni^{2+}5.0, Zn^{2+} 3.5. biomass concentration 3g/l,particle size 300-600 μm.

3.2 Influence of pH

The pH is an important parameter for adsorption of heavy metal ions. Under acidic condition, little sorption of the metal ions was found. P.R.Purank (P.R.Punrank) also confirmed the similar results. When the pH is high (>7.0), some metal ions will precipitate. When the pH is higher than 5.0, which is near to the pKa of chitosan(pH5.3-5.5) , the high adsorption capacity could be obtained.The optimal pH for sorption of metal ions are following: Cu^{2+}4.5 to 5.5, Ni^{2+}:5.0 to 7.0; Zn^{2+}:4.0 to 6.0.

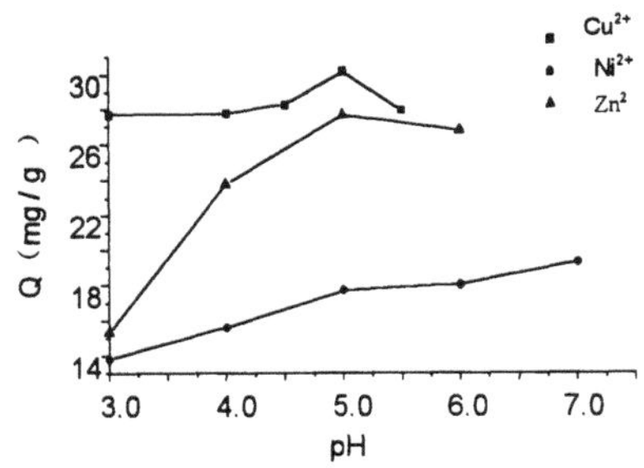

Fig.2 Effect of pH on adsorption
biomass concentration 3g/l,paticle size 300-600μm.

3.3 Effect of particle size

For normal ion exchange resin, the smaller the particle, the higher sorption capacity could be obtained. For the biomass of penicillin chromyceall, small particle (200-300 μm) had higher capacity, but the influence seems not so considerable(Fig.3).

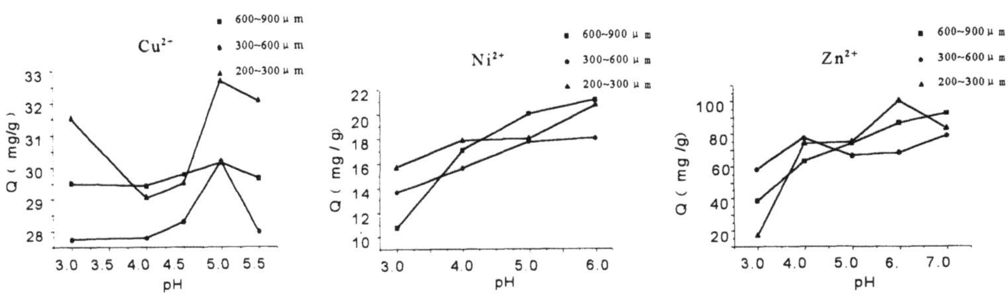

Fig. 3 Effect of particle size on adsorption

3.4 Effect of salt concentration

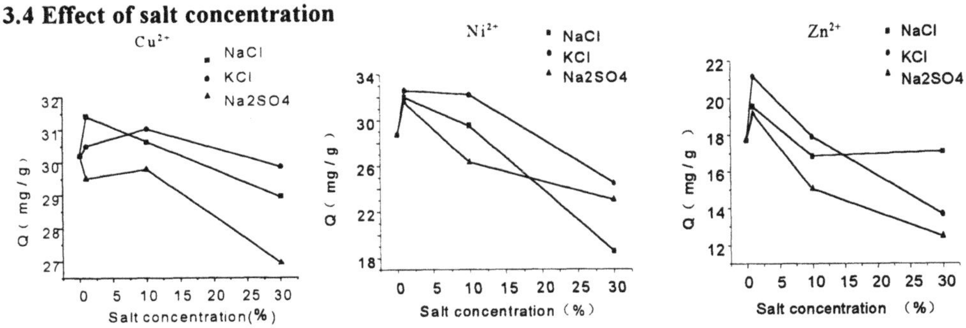

Fig.4. Effect of salt concentrati on on adsorption

Many salts such as NaCl, often existed in waste water and these salts will influence up take of metal ions. It is found that low salt concentration enhanced the sorption capacity. High salt concentration would decrease sorption of metal ions because high salt ion willl inhibit the sorption of metal ion to the biomass. It is not clearly now why low salt concentration increases the up take of metal ions.

3.5 Sorption model

Langmuir Model and Freundlish Model are often used to describe the equilibrium sorption isotherms

Langmuir Model: Ceq/Q=1/(bQmax)+Ceq/Qmax

Ceq is metal ion concentration at equilibrium; Q and Qmax is metal up take and max metal ion up take(mg/g); b is Langmuir isotherm constant.

Freundlish Model: logQ=logK+1/n logCeq

K and n is the constant , respectively.

The simulated results with comparison with experiment date were shown in Fig.5 and Fig.6.

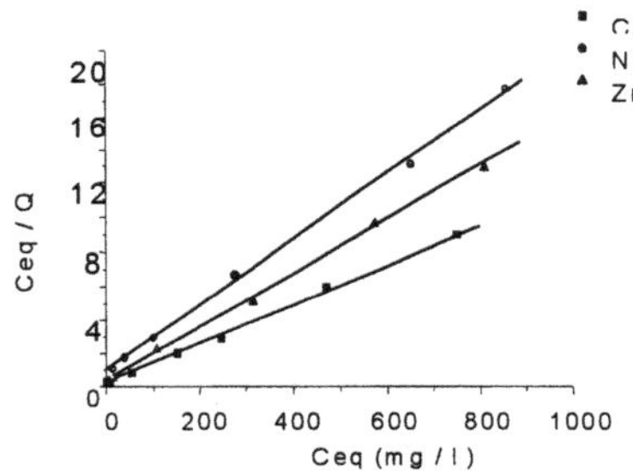

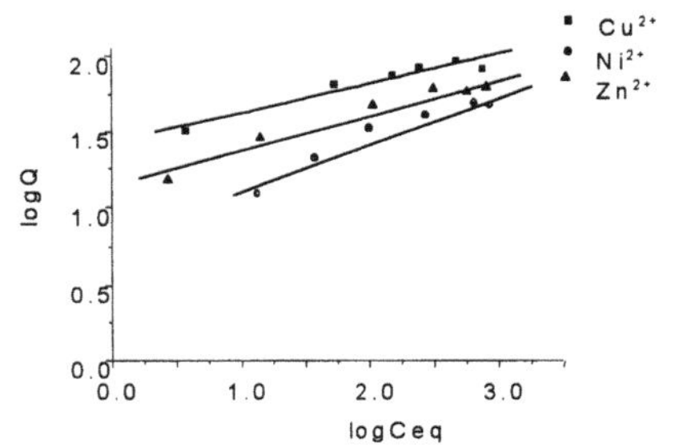

Fig.5 Simulated result by Langmuir model Fig.6 Simulated result by Freundlish model

The two models can be used to describe the sorption isotherms of metal ions, which indicated that the biosorption of metals can be a mono lay sorption without interaction between adsorbed metal ions.

3.6 Desorption

From above results, the desorption could be obtained by decreasing pH . The desorption results are shown in Table 1

Table1 Desorption of metal ion

Desorption condition	0.5M HCl	0.5M HNO_3	0.5M citric acid	0.1M NaCl	0. 02M EDTA
Desorption %	85	95	86	30	52

0.5M HNO_3 was used to desorb the metal ion. After desorption, the resin could be reused at least 3 times without losing adsorption capacity considerabely.

3.7 Comparison with other ion exchange resins

The comparison of the penicillin biomass adsorbent with CM—Sephadex C-25 and chelated resin D-751 was made. The adsorption capacity of heavy metal is shown in following table 2

Table 2 Comparison of different adsorbents

Adsorbed capacity	Penicillin mycelium	Chelated D-751	Prototype Gel	CM-Sephadex C-25
Cu^{2+} (mg/ g)	93.4	170	48.3	170
Zn^{2+} (mg/g)	63	187	28.4	159
Cr^{3+} (mg/g)	97	99	-----	132
Ni^{2+} (mg/g)	49	148	53.73	124

Despite the CM-Sephadex C-25 and Chelated D-751 resin have high adsorbed capacity than that of penicillin biomass. The penicillin mycelium has low cost and the adsorbed heavy metal ions are easily desorbed which has potential in large-scale removal of heavy metal ions in waster water treatment.

CONCLUSION

The waste biomass from penicillin industry has high sorption capacity for many heavy metal ions. The biomass could be used in waste water treatment due to low cost and high selectivity.The sorption was influenced by many parameters such as pH, salt concentration and so on.The isotherms could be described by Freundhlish and Langmuir models. The desorption could be achieved by 0.5 M HNO3 and the biomass could be :eused many times.

REFERENCES

1. Holan,Z.R., Volesky, B., Prasetyo, I., Bioseportion of cadmium by biomass of marine algae. Biotechnol.Bioeng. 4(1993),819-825
2. Zhu ,Y., Sengupta A.K., Sorption Enhancement of Some hydrophilic Organic Solutes through Polymeric Ligand Exchange, Environ.Sci.Technol., Vol26, No.10(1992):1990-1997
3. Puranik, K.M.Paknikar, Bioseparation of lead and zinc from solutions using Streptoverticillium cinnamoneum Waster biomass, Journal of Biotechnology, 55(1997):113-124
4. Fourest, E., Roux,J.C., Heavy metal biosorption by fungal mycelial by products:mechanisms and influence of pH:App.Microbiol.Biotechnol.37(1992):399-403
5. Ma, B.X, Photometric analysis of Cu^{2-} with BCO, Chinses Journal of Analtical Chemistry,1974,2(2):53-56
6. Luo, Z.M., Ceng, P.R., Reaction between Zincon and Zn with Tween-80 existent and its applications,Chinese Journal of Physical and Chemical Measurement,1989 25(5):274-276

Bioseparation Engineering
I. Endo, T. Nagamune, S. Katoh and T. Yonemoto (Editors)

Removal of Cadmium Ion by the Moss *Pohlia flexuosa*

M. Azuma, A. Obayashi, M. Kondoh, C. Kawasaki, K. Igarashi, J. Kato, and H. Ooshima

Department of Bioapplied Chemistry, Osaka City University, Japan

We examined cadmium removal from aqueous solution using the moss *Pohlia flexuosa*. The amounts of cadmium ions removed increased with increasing initial concentration of the ion over the range from 10 to 100 μg/ml. The concentration factor, which is an indicator of the ability to remove heavy metals, was 770 ml/g of dry moss at 10 μg/ml of $CdCl_2$. We examined the effects of temperature, respiratory inhibitors, and pH on the cadmium removal from 30 and 100 μg/ml $CdCl_2$ solution. The results suggested that cadmium removal at these concentrations may be mainly dependent on physical adsorption on the cell surface. The carboxyl groups of the moss were modified by the HCl / methanol esterification method. The modified moss showed almost no cadmium ions removal. Carboxyl groups on the cell surface may markedly contribute to the adsorption. We discussed the process of cadmium removal from solution using the moss in comparison with the process of mercury removal, which was reported previously.

(Key words: Bioremoval, Moss, *Pohlia flexuosa*, Cadmium removal, Carboxyl groups)

1. INTRODUCTION

Toxic heavy metal contamination of the environment has become a very significant problem. Biological systems have been suggested to be adapted for removal of toxic heavy metal from industrial aqueous waste. Bioremoval has been studied using bacteria, yeast, fungi, microalgae and seaweed (1-5). However, it is still not practicable for industrial use. On the other hand, mosses have been used as a indicators of heavy metals contamination because they are very sensitive to heavy metals (6-9). Previously, we reported Hg^{2+} removal using the moss *Pohlia flexuosa* (10). The moss was obtained as the protonema from its matured capsules, and was grown by shaking in modified Murashige-Skoog's medium. The mature protonema removed Hg^{2+} from aqueous solutions. The amounts of Hg^{2+} removed increased with increasing initial Hg^{2+} concentration over the range from 10 to 100 μg/ml. The concentration factor, which is an indicator of the ability to remove heavy metals, was 2000 ml/g of dry moss at 50 μg/ml of $HgCl_2$. This value was similar to the known concentration factor for Cd^{2+} for some cyanobacteria and for some algae. This result suggested that the moss is a biomasss suitable for the removal of toxic metals. Also, the effects of temperature, pH and respiratory inhibitors on Hg^{2+} removal were examined. Based on these results the importance of two processes, bioaccumulation and adsorption, in the Hg^{2+} removal was discussed. Here, the removal of Cd^{2+} using the moss was investigated, and the process of Cd^{2+} removal will be discussed.

2. METHODS

2.1 Cd^{2+} removal from solution using the moss

The moss was cultured under the conditions reported previously (10). Cd^{2+} removal from solution was carried out as follows. A given amount of the moss (15 – 25 mg dry weight) was transferred to 20 ml of a solution (approximately pH 6) containing 10 – 100 μg/ml $CdCl_2$. The solution containing protonema was shaken on a rotary shaker at 25 °C and aliquots of the solutions were taken for analysis after a given period of incubation. The biomass was separated using a 0.45 μm filter and the filtrate was analyzed for Cd^{2+} using an inductively coupled plasma atomic emission spectroscope (ICPS-1000TR, Shimadzu Co., Kyoto). Cd^{2+} removal was defined as (C_0-C) / mg dry weight of biomass, where C_0 and C represent the $CdCl_2$ concentration in the aqueous phase before and after contact with the biomass, respectively. Autoclave treatment of the moss was carried out at 121 °C for 10 min.

2.2 Cd^{2+} release from the moss

The moss was suspended in Cd^{2+} solution (100 μg/ml) for 120 min, as described above, recovered by filtration, and washed quickly with 20 ml of deionized water (pH 6). About 100 mg dry weight of the moss was suspended for 120 min with 20 ml of 10 mM phosphate buffer in the pH range of 3.5 to 6.0, and separated using a 0.45 μm filter. The filtrate was analyzed using the ICPS-1000TR. The amount of Cd^{2+} release (%) was determined on the basis of Cd^{2+} amount in the moss prior to suspending with phosphate buffer.

2.3 Modification of carboxyl groups

HCl / methanol esterification performed according to the method reported previously to modify carboxyl groups of fungal cell wall (11-12). The moss was cultured and the remaining water on the cell surface was removed using filters. Aliquots of 2 g of the moss were suspended in 300 ml of methanol (99.9 %) and 2.5 ml of 12N HCl (0.1 M HCl final concentration) , with continuous agitation. After a given period of incubation, the mosses were separated using a 0.45 μm filter, and then washed quickly twice with deionized water. This moss was used for the Cd^{2+} removal experiment.

The degree of esterification of the moss was examined by alkaline hydrolysis (12). Analysis of the methanol released from the modified moss was performed by Gas Chromatography (GC-8A, Shimadzu) under the following conditions: column, Gasukuropack 56 (GL Science Inc.); Injector temperature, 180 °C; column temperature, 150 °C; carrier gas,

N_2; and detector, FID. The solution for analysis of the released methanol contained the remaining methanol in the moss because the moss was contact with the HCl / methanol solution. Therefore, degree of esterification degree was defined as (released methanol at each time point) - (released methanol at zero time).

3. RESULTS AND DISCUSSION

3.1 Removal of Cd^{2+} from solution

The efficiency of Cd^{2+} removal from aqueous solution using the moss was investigated. The amount of Cd^{2+} removal increased with increasing initial Cd^{2+} concentration over the range from 10 to 100 μg/ml (Fig. 1). In all cases, Cd^{2+} removal increased sharply until 10 min of exposure and then increased slowly. In initial Cd^{2+} concentrations of 10 and 25 μg/ml, Cd^{2+} removal was almost constant between exposure time of 60 and 120 min. However, at 50 and 100 μg/ml Cd^{2+} removal increased slightly between 60 and 120 min. At 120 min of incubation, the ratios of Cd^{2+} removal at initial Cd^{2+} concentrations of 10, 25, 50 and 100 μg/ml were about 92, 81, 73 and 49 %, respectively. This percent removal was defined as (initial Cd^{2+} conc. - Cd^{2+} conc.) / (initial Cd^{2+} conc.) x 100. The ability of biomass for the toxic metal removal is generally expressed as the concentration factor, which is the ratio of (μg metal removed / g dry weight) to (μg metal / ml of solution). The concentration factor was about 770 ml/g of dry moss at 10 μg/ml of $CdCl_2$, and was 610 ml/g of dry moss at 50 μg/ml of $CdCl_2$ where the factor was calculated from the value after 120 min of exposure although Cd^{2+} removal had not yet reached equilibrium at this time point. These values were smaller

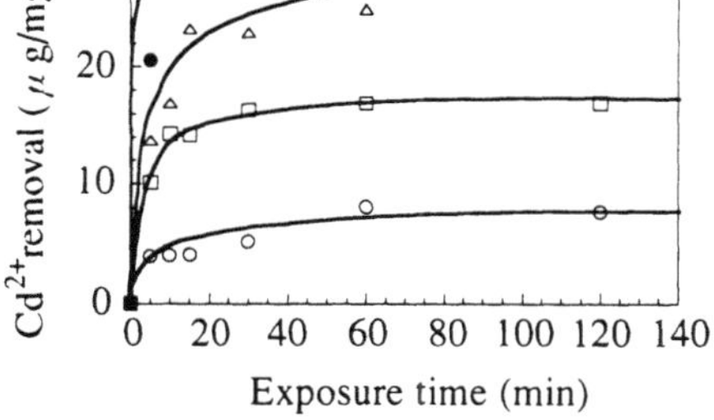

Fig. 1 Effects of initial concentration of cadmium ions on cadmium removal.

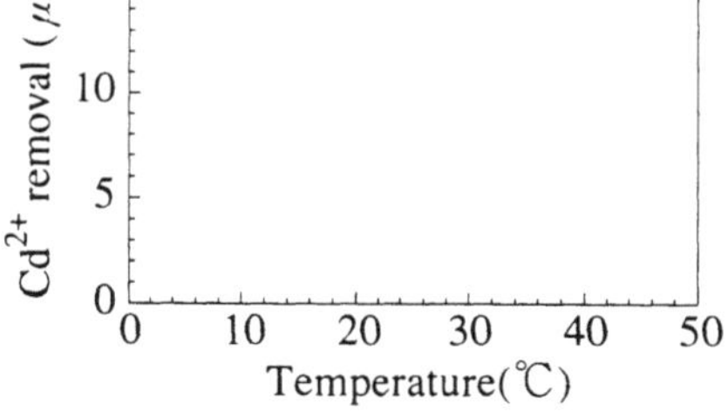

Fig. 2 Effects of temperature on cadmium removal. Initial cadmium conc.: 30 μg/ml. Exposure time: 120 min.

than the value of Hg^{2+} removal (2000 ml/g of dry moss at 50 μg/ml of $HgCl_2$) (10). The ability of the moss to remove Cd^{2+} was lower than that for Hg^{2+} removal.

3.2 Effects of temperature, respiratory inhibitors and pH on Cd^{2+} removal

We previously reported that optimum pH in Hg^{2+} removal by the moss was 8.0, that the optimum temperature was 25 °C, and that addition of respiratory inhibitors partially inhibited Hg^{2+} removal. These results suggested that Hg^{2+} removal by the moss involves two processes; adsorption on the cell surface, and uptake into the cells.

We examined the effects of pH and temperature on Cd^{2+} removal. The effect of temperature was shown in Fig. 2. Cd^{2+} removal was almost constant over the temperature range from 4 to 15 °C, and then increased slightly with rising temperature from 15 to 45 °C. The temperature dependency of Cd^{2+} removal was different from that of Hg^{2+} removal. The effects of temperature on Cd^{2+} removal by autoclaved and untreated moss were the same. Thus, there were no differences in Cd^{2+} removal between living and dead moss. These results suggested that Cd^{2+} removal from solutions containing 30 μg/ml $CdCl_2$ may be dependent on adsorption onto the cell surface.

In a subsequent experiment, we examined the effects of respiratory inhibitors on Cd^{2+} removal (Table 1). Addition of sodium azide (0.1 – 10 mM) and dinitrophenol (DNP, 0.1 and 1.0 mM), respiratory inhibitors, showed almost no effect on Cd^{2+} removal by the moss. In contrast, these concentrations of inhibitors blocked Hg^{2+} removal (10). These results were consistent with the above suggestion obtained from the effects of temperature.

Table 1 Effects of respiratory inhibitors on cadmium removal.

	Concentration (mM)	Cadmium removal (μg/mg dry weight)	Relative removal (%)
None		58.6	100
NaN_3	0.1	60.9	104
	1.0	62.6	107
	10.0	56.4	96
DNP	0.1	51.5	88
	1.0	58.5	100

Initial cadmium conc. 100 μg/ml. Exposure time 120min.

The effect of pH is shown in Fig. 3. The removal increased sharply with increasing pH over the range from 3 to 6, and then remained almost constant between 6 and 10. The pH profile was different from that observed for Hg^{2+} removal. In a subsequent experiment, we examined the effects of pH on Cd^{2+} release from the moss (Fig. 4). After contact with Cd^{2+} solution (100 μg/ml), the moss was suspended in 10 mM phosphate buffer (20 ml) at pH ranging from 3.5 to 6.0 for 120 min. The Cd^{2+} release increased with decreasing pH over the range from 3.5 to 6. At pH 3.5 and 6.0, the amounts of Cd^{2+} release (%) were about 26 and

5 %, respectively. The effects of pH on Cd^{2+} removal from solution and Cd^{2+} release from the moss, especially behavior at pH 3.5 and below, suggested that carboxyl groups on the cell surface of the moss may greatly contribute to Cd^{2+} removal. Our results suggested that the process of Cd^{2+} removal was dependent on adsorption onto the cell surface. Mosses are very sensitive to heavy metals (6-9), and in our preliminary experiment the moss did not grow for 10 days after coming into contact with Cd^{2+} solution (10 – 100 μg/ml of $CdCl_2$) for only 120 min. Cd^{2+} concentration may have been too high to observe its uptake into cells.

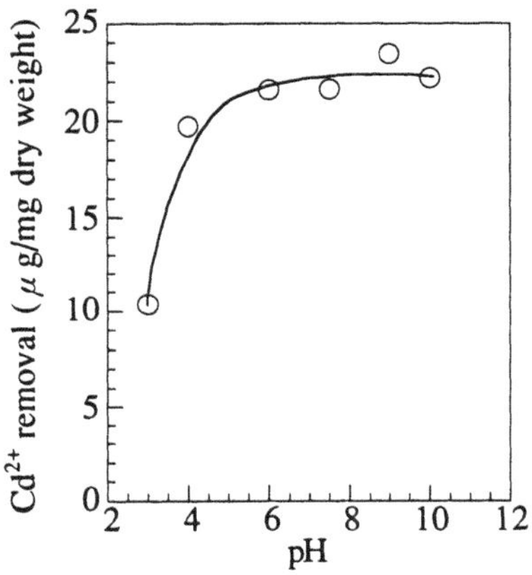

Fig. 3 Effects of pH on cadmium removal from solution.
Initial cadmium conc.: 30 μ g/ml.
Exposure time: 120 min.

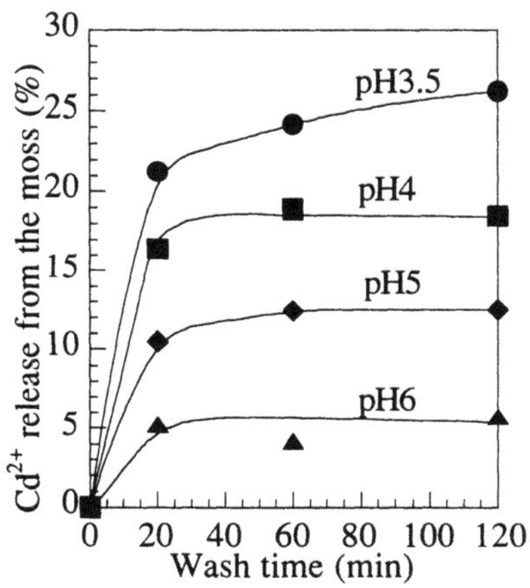

Fig. 4 Effects of pH on cadmium release from the moss.

3.3 Effects of modification of carboxyl groups on Cd^{2+} removal

The effects of modification of carboxyl groups by the HCl / methanol method on Cd^{2+} removal were investigated. The moss was first treated with methanol, HCl or methanol / HCl for 48h, then Cd^{2+} removal experiments were carried out. Methanol treatment of the moss led to a slight increase of Cd^{2+} removal in comparison with untreated moss, and acidic treatment (0.1 N HCl) led to a decrease to approximately 50 %. This decrease may be because the solution used in the Cd^{2+} removal experiment was acidic due to the remaining HCl in the moss. Fig. 5 shows carboxyl esterification kinetics by acidic methanol treatment. The Cd^{2+} removal after treatment for 48 h was decreased to approximately 8 % of that in untreated. Degree of esterification degree was examined by analysis of methanol released from the modified moss. The degree increased with esterification reaction time. The Cd^{2+} removal decreased with increasing degree of esterification. These results suggested that carboxyl groups of the moss may contribute to a large extent to cadmium binding.

In conclusion, we found that the Cd^{2+} removal ability of the moss was lower than that for

Hg^{2+} removal, that Cd^{2+} removal may be almost independent of the uptake of Cd^{2+} into the cells in the presence of relatively high concentrations of Cd^{2+}, and that carboxyl groups of the moss may contribute to a large extent to cadmium binding.

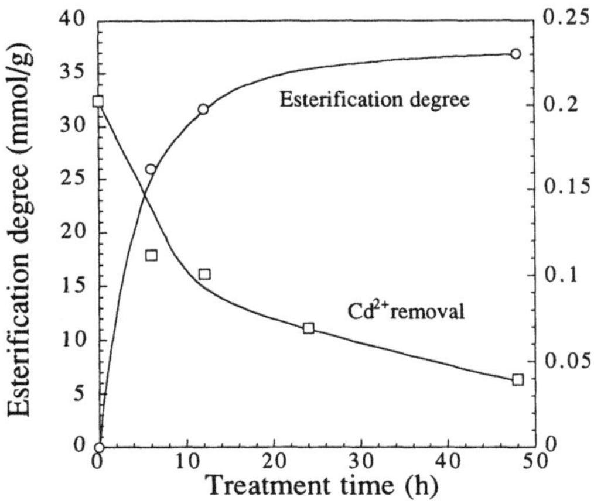

Fig. 5 Carboxyl esterification kinetics by acidic methanol treatment.

References

(1) Buelva, L. B., Kakii, K., and Kuriyama, M., J Ferment. Bioeng., 79 (1995)136.

(2) Voresky, B. and May-Phillips, H. A., Appl. Microbiol. Biotechnol., 42 (1995) 797.

(3) Al-Asheh, S. and Duvnjak, Z., Biotechnol. Prog., 11 (1995) 638.

(4) Inthorn, D., Nagase, H., Isaji, Y., Hiratas, K., and Miyamoto, K., J Ferment. Bioeng., 82 (1996)580.

(5) Coasta, A. C. A. and Faranca, F. P., Sep. Sci. Technol., 31 (1996) 2373.

(6) Beckett, P. J., Environ. Contam., Int. Conf. 2nd, (1986) 30.

(7) Erdman, J. A. and Modreski, P. J., J. Geochem. Explor, 20 (1984) 75.

(8) Beaugelin-Seiller, K., Baudin, J. P., and Casellas, C., Arch. Environ. Contam. Toxicol., 28, (1995) 125.

(9) Frontasyeva, M. V. and Steinnes, E., Analys., 120 (1995) 1437.

(10) Kondoh, M., Fukuda, M., Azuma, M., Ooshima, H., and Kato, J., J. Ferment. Bioeng., 86 (1998) 197.

(11) Fourest, E., Serre, A., and Roux, J.-C., Toxicol. Environ. Chem., 54 (1996) 1.

(12) Gardea-Torresday, J. L., Becker-Hapak, M. K., Hosea, J. M., and Darnall, D. W., Environ. Sci. Technol., 24 (1990) 1372.

Bioseparation Engineering
I. Endo, T. Nagamune, S. Katoh and T. Yonemoto (Editors)

The effects of additives on hydrolysis of cellulose with water under pressures

T. Funazukuri[a], M. Hirota[a], T. Nagatake[a], and M. Goto[b]

[a] Department of Applied Chemistry, Chuo University, 1-13-27 Kasuga, Bunkyo-ku, Tokyo 112-8551, Japan,

[b] Department of Applied Chemistry and Biochemistry, Kumamoto University, 2-39-1 Kurokami, Kumamoto, 860-8555, Japan

The effects of additives on hydrolysis rate and glucose yield were studied with bomb-type small batch reactors in hydrolysis of cellulosic samples under pressures in the temperature range from 250℃ to 380 ℃. Although the rates were significantly enhanced by acidic additives (H_2SO_4, CH_3COOH), the glucose yields were not. The yields with water were higher than those with water+ additives.

1. INTRODUCTION

A large amount of cellulosic materials involved in municipal solid wastes have been disposed of by incineration and landfill, although some of them are utilized as recycled paper, fuel, packing materials etc. In order to decrease the environmental impact and increase the recycle rate of cellulosic wastes, the useful conversion process for cellulose to valuable materials has been required. One of the promising processes is hydrolysis of cellulose followed by fermentation to produce ethanol. A large number of studies on hydrolysis of cellulose have been made with various acids or enzymes at low pressures. Recently, Adschiri et al. (1) reported that cellulose was effectively hydrolyzed by contacting subcritical or supercritical water without any additives at extremely short residence time. The results are very attractive, but the engineering problems in both the feed of the sample and the attainability of short residence times have arisen. In this study in order to overcome these, the reaction temperature is decreased by adding the small amount of acids, and then the residence times are increased. The effects of the additives on cellulose conversion and glucose yields are studied.

2. EXPERIMENTAL APPARATUS AND PROCEDURES

A batch tubular reactor (5.3 ml), made of stainless steel, was used. The both ends were plugged with Swagelok caps, and sometimes one end was connected to a valve via 1/16-inch tubing. Asheless filter paper (TOYO No.2, Advantech) and microcrystalline cellulose powers (Avicel, Merck and Advantech) were employed. At room temperature the cellulosic sample (0.2-0.3 mg), and the certain amount (3.0-5.0ml) of distilled water or aqueous solution were loaded. The air in the reactor was purged with argon, and then it was plugged. At time zero the reactor was immersed in the salt-bath, equipped with a stirrer, whose temperature had been maintained at the prescribed temperature. After certain time elapsed, the reactor was removed from the bath and cooled by water. The reactor was opened, and then the solid sample and the reaction solvent were recovered. The inside reactor was washed with distilled water. The reaction solvent and the washing solvent were mixed and filtered. The solid sample was dried and weighed. The products in the filtrate were mainly analyzed by HPLC (Sugar KS801, Shodex). The silylated products were also analyzed by a GC-MS

3. RESULTS AND DISCUSSION

As reported by many workers, the hydrolysis rates of cellulosic samples are assumed to be represented by the first order reaction kinetics, given by eq (1)

$$-\frac{dW}{dt} = k\left(W - W_\infty\right) \qquad (1)$$

where W and W_∞ are sample weight at reaction time t and solid residue at hydrolysis completion, respectively, and k is the rate constant. For pure cellulose W_∞ can be assumed to be zero.

Figure 1 shows $\ln(W/W_0)$ vs. time for the filter paper with water in the absence of additives. The reaction time was counted from the time the reactor was immersed in the salt bath. The data at each temperature are represented by each straight line while some data are scattered. The overall hydrolysis rates in terms of sample weight change are the first order reaction kinetics. The actual temperature changes were measured inside the reactor with a thermocouple placed in the reactor, and the heat-up period of nearly 0.5 min to reach the prescribed temperature was observed. This time delay was taken into account to obtain the rates.

Figure 2 shows yields of the precipitated product vs. time for the filter paper with water. The solid products were precipitated after the product solution was decanted. This white fine particles probably seem to be not char but depolymerized cellulosic sample. The maximum yields at higher temperatures are found to be shorter reaction times. The maximum yields at 260 ℃ for 5 min and 240 ℃ for 20 min attain to about 90 % of the sample, and the yields decrease markedly at longer reaction times.

Figures 3 and **4** show glucose yields from the filter paper with water on basis of initial weight of the sample loaded and the reacted sample weight, respectively. The maximum glucose yields are as high as 10% of the initial sample weight at 260℃ for 20 min. At longer reaction times the yields decrease considerably. Almost all of the reacted cellulose is converted to glucose under this condition.

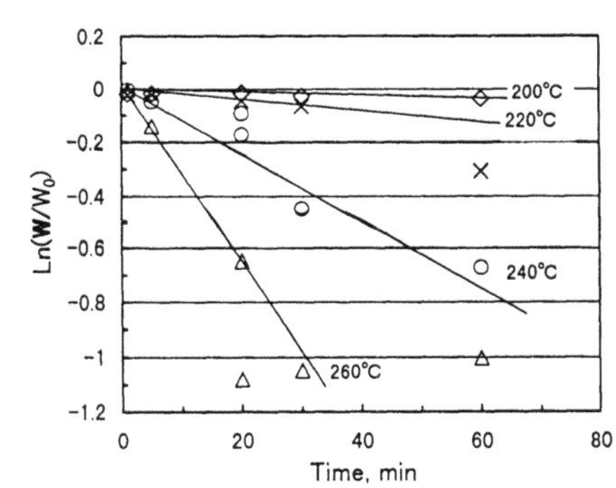

Fig. 1 $\ln(W/W_0)$ vs. time for filter paper with water.

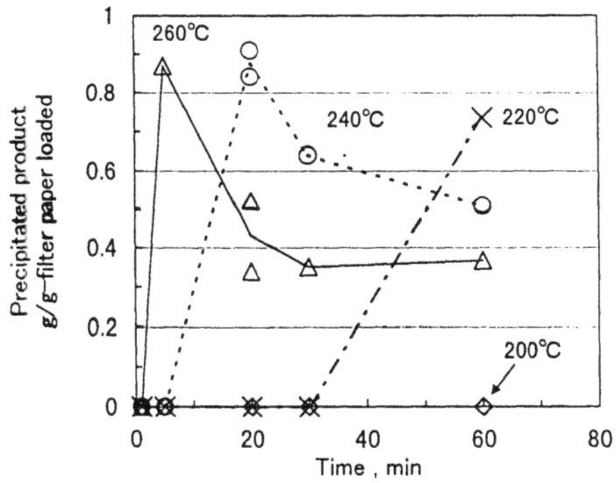

Fig. 2 Yield of precipitated product vs. time for filter paper with water.

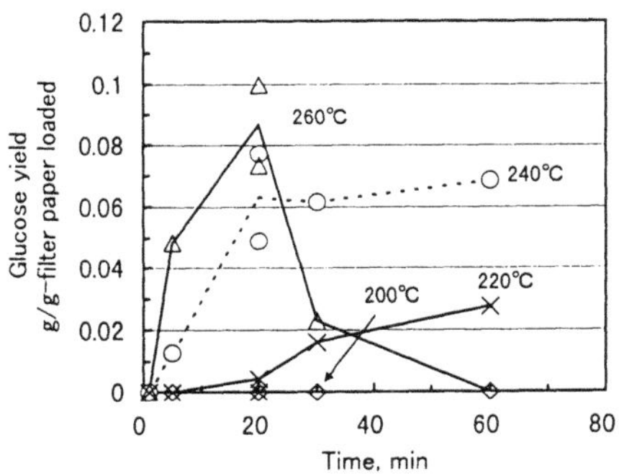

Fig. 3 Glucose yield vs. time for filter paper with water.

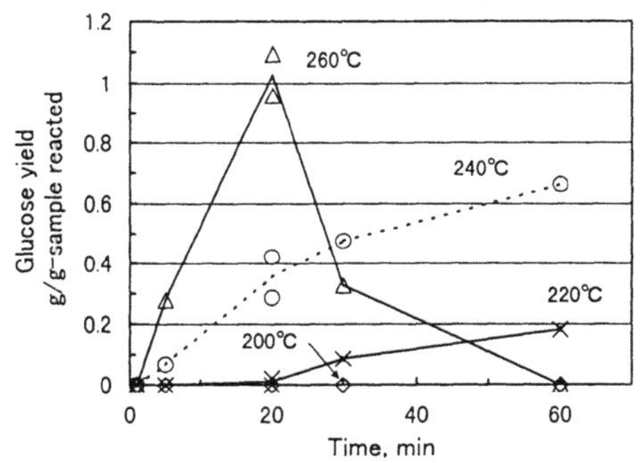

Fig. 4 Glucose yield vs. time for filter paper with water, based on sample reacted.

Note that yield of the glucose was analyzed with HPLC, and the separation was not perfect. Then the yields involve those of 1,6-anhydro- β -D-glucose possibly.

Figure 5 shows the sample weight loss for the filter paper with various additives at 240℃. The 0.0001N-H_2SO_4 aqueous solution is found to hardly affect the sample weight loss, but the 0.01N-H_2SO_4 aqueous solution increases sample weight loss. The addition of CH_3COOH in the small amount affects the weight loss slightly, but the neat of CH_3COOH does not affect the weight loss at longer times due to acetylation.

Figure 6 shows the effects of additives on glucose yields at 240℃. The maximum glucose yields in terms of sample reacted are observed at 10 min, and no additive (pure water) is the most effective on the glucose production.

Figure 7 shows the Arrhenius plots of overall hydrolysis rate constants of

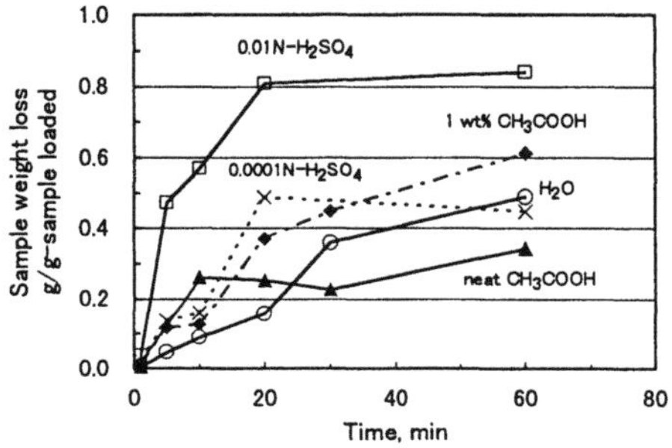

Fig. 5 Sample weight loss vs. time for filter paper with/without additives at 240 ℃.

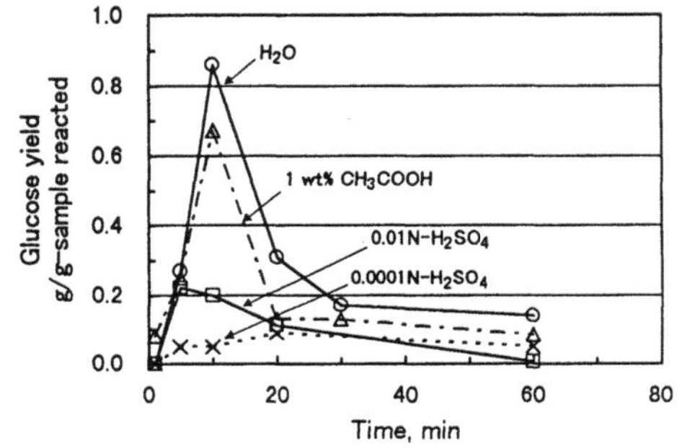

Fig. 6 Glucose yield vs. time for filter paper with/without additives at 240 ℃.

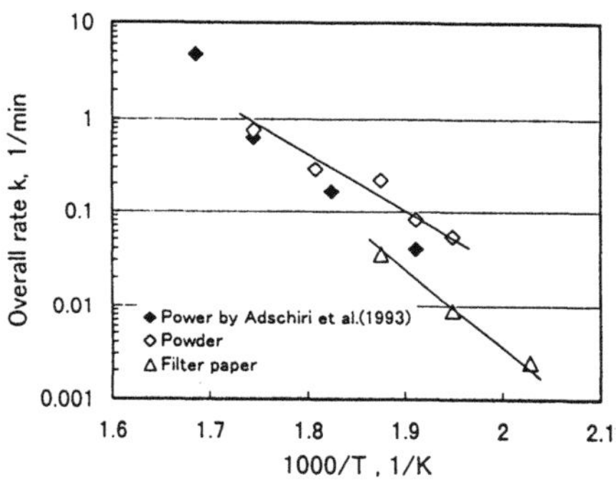

Fig. 7 Arrhenius plot for cellulose powder and filter paper, cpmpared with those of Adschiri et al.(1993)

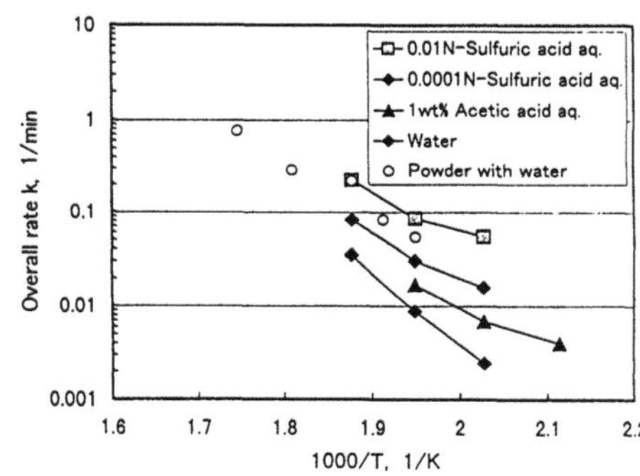

Fig. 8 Arrhenius plot for filter paper with various additives

microcrystalline cellulose powder and filter paper in the absence of the additives in this study, together with those of cellulose powder by Adschiri et al.(1993). Although the rates of the powder at lower temperatures are slightly higher than those of Adschiri et al., the both values become consistent at higher temperatures.

Figure 8 shows the Arrhenius plots for overall hydrolysis rates of the filter paper in the presence of the acidic additives. The rates increase in the order water to 1-wt% acetic acid aq. to 0.0001-N sulfuric acid aq. to 0.01-N sulfuric acid aq. All the plots are not represented with straight lines, and the values at higher temperatures are higher than the straight lines. This may be considered to result from the effect of thermal decomposition, as pointed out by Adschiri et al.

Figure 9 shows glucose yields from hydrolysis of cellulose powder (Advantech) with water. The maximum yield is obtained at 270℃ for 5 min excluding heat-up

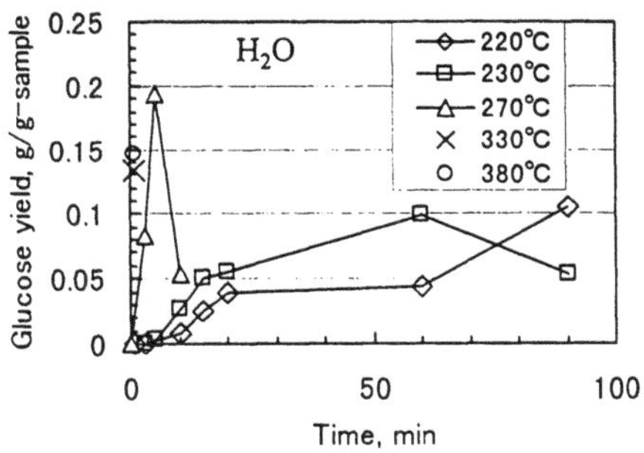

Fig. 9 Glucose yields from cellulose powder with water at various temperatures.

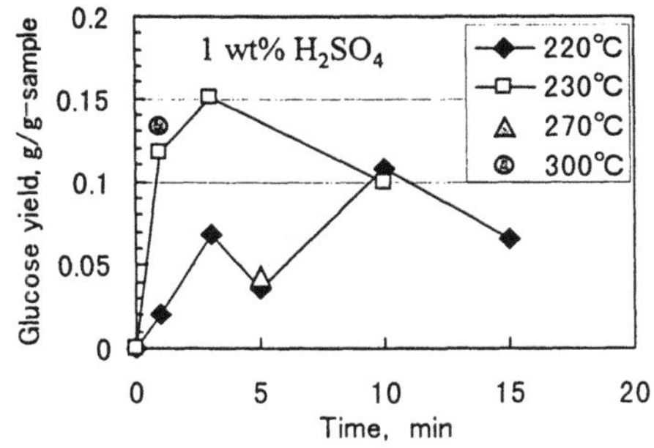

Fig. 10 Glucose yields from cellulose powder with 1-wt% H_2SO_4 aq. solution at various temperatures.

period of 0.5 min, but the yield reaches 15-% of initial sample weight at 380℃ for 0.5 min. **Figure 10** shows the yields with 1-wt% H_2SO_4 aqueous solutions. The maximum glucose yield of 15% of initial sample weight is obtained at 270℃ for 3 min, and almost the same yield is that at 300℃ and 0.5 min.

4. CONCLUSIONS

The weight losses of hydrolysis of filter paper and cellulose powder were expressed with the first order reaction kinetics with and without additives. The rate constants for cellulose powder with water at higher temperatures were almost consistent with those of Adschiri et al., and increased with the additives. The hydrolysis rate for cellulose powder was higher than that for filter paper. It was found that glucose was mainly produced in the region of low cellulose conversions, but the yields did not increase with the conversions. The glucose yields with water, on the basis of sample reacted, were higher than those with the acidic additives.

ACKNOWLEDGEMENT

The authors were grateful to the New Energy and Industrial Development Organization for the International Joint Research Grant.

REFERENCES

(1) Adschiri, T., Hirose, S., Malaluan, R. and Arai, K., "Noncatalytic conversion of cellulose in supercritical and subcritical water", J. Chem. Eng. Japan, **26**, 676-680 (1993).

Key words: Hydrolysis, Cellulose, Rate, Glucose, High pressure

Bioseparation Engineering
I. Endo, T. Nagamune, S. Katoh and T. Yonemoto (Editors)

Removal of volatile organic compounds from waste gas in packed column with immobilized activated sludge gel beads.

K. Nakao, M. A. Ibrahim, Y. Yasuda and K. Fukunaga

Department of Applied Chemistry and Chemical Engineering, Yamaguchi University, 2-16-1 Tokiwadai, Ube, Yamaguchi, 755-8611, Japan.

The packed column bioreactor containing the immobilized activated sludge gel beads as well as hollow plastic balls to avoid compaction of gel beads was used to remove toluene, benzene or ethylbenzene as a model of the toxic volatile organic compounds (VOC). The basic process parameters such as the specific rate constant of the VOC degradation by the gel beads, the Henry's law constant of the VOC etc. were firstly determined. For a wide range of VOC concentration, the Monod-type degradation rate equation was assumed. Secondly, the removals observed in the packed column were compared with the ones calculated using the proposed design equation with the parameters determined. The packed column was efficient for one year to remove the VOC and its removal performance could be predicted by the design equation.

KEYWORDS : Immobilized activated sludge, Hollow plastic balls, VOC biodegradation, Packed column, Waste gas treatment

1. INTRODUCTION

The biological waste gas treatment is superior to the conventional physical or chemical processes due to its wider applicability, no secondary pollution and saving in both energy and chemicals(Nakao et al.,1990). Various types of porous materials have been employed as biofilter media for treatment of waste gas containing malodorant and toxic volatile organic compound(VOC). Compost, peat, wood chip and soil are the commonly used media, on which microorganisms are immobilized to form biolayer(William and Miller.,1992). However, these natural organic porous media have some drawbacks such as progressive compaction, channeling and biodegradation (Kirchner et al.,1989; Hartmans and Tramper, 1991; Togna and Folsom, 1992; Sorial et al.,1993; Smith et al., 1993). In the previous work(Nakao et al.,1990), an acclimated activated sludge was entrapped in stable synthetic polymer networks of gel beads in which polyvinyl alcohol was cross-linked by boric acid. None of other studies has examined the potential benefit of immobilized activated sludge gel beads to degrade air pollutant. Thus, the biofiltration process employing the packed column which contained the immobilized activated sludge gel beads and the hollow plastic balls was used to remove acetaldehyde and propionaldehyde(Nakao et al., 1994). Then, the research was extended to treat VOC such as toluene, and benzene(Ibrahim et al., 1997). These compounds are hazardous and may cause cancer to human being. The gel beads used were the same as those in the previous work(Nakao et al., 1994) and acclimated with the VOC in the packed column. The objective of this work is to asses the potential of the immobilized activated sludge gel

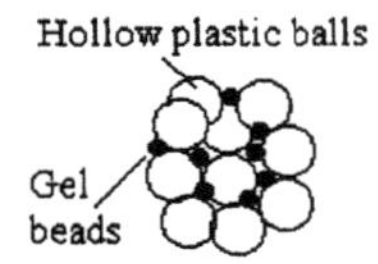

(a)

Figure 1(a) Avoidance of gel beads compaction by hollow plastic balls

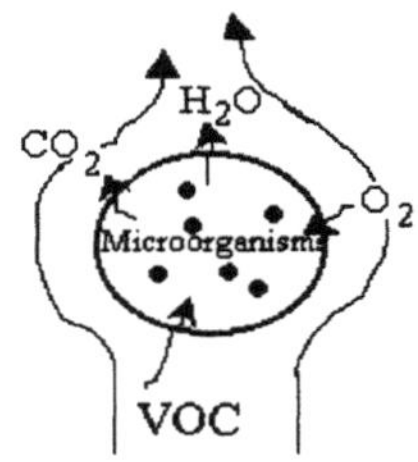

(b)

Figure 1(b) Biodegradation within gel beads

beads for removal of VOCs (toluene, benzene and ethylbenzene) in the packed column and to establish an optimal design and operation for the waste gas treatment process. For this purpose, it is firstly essential to determine the basic process parameters such as the specific rate constant of VOC degradation by the gel beads, the Henry's law constant of VOC etc. For a wide range of VOC concentration, the Monod-type degradation rate equation is assumed. Secondly, the removals observed in the packed column are compared with the ones calculated using the proposed design equation with the parameters determined.

2. THEORY

The waste gas flows through beds of the immobilized activated sludge gel beads and the hollow plastic balls to avoid compaction of gel beads and hence reduce pressure drop for gas flow as described in **Figure 1(a)**. A pollutant(VOC) in the gas stream is transferred to the surface of the gel bead, dissolved in water at the surface and transferred inward the gel bead while being degraded by microorganisms in activated sludge entrapped in the gel bead as indicated in **Figure 1(b)**. No mass transfer limitation is assumed to exist not only between the gas phase and the gel beads surface but also between the surface and the entrapped activated sludge since the biodegradation rate is much lower than the transfer rate. The kinetics of the VOC degradation by microorganisms in the gel beads is assumed to follow the Monod type relationship as shown in **Table 1**. The design equation for the VOC removal operation in the gel beads packed column can be derived based on the Monod type rate equation as shown in **Table 2**.

Table 1
Monod type biodegradation rate equation

$$-r_A = \frac{-dC_A}{dt} = \frac{k_1 C_A X}{1+k_2 C_A}\left(\frac{V_g}{V_g+V_l}\right) \qquad (1)$$

Integrating Eq.(1)

$$\ln\frac{\left(\frac{C_{A,O}}{C_A}\right)}{\left(C_{A,O}-C_A\right)} + k_2 = k_1 X\left(\frac{V_g}{V_r}\right)\frac{t}{\left(C_{A,O}-C_A\right)} \qquad (2)$$

where, $C_{A,o}$: Initial concentration of VOC in deionized water [mg/l], C_A: Concentration of VOC in deionized water [mg/l], X : Activated sludge concentration in the gel beads[kgMLSS/ m^3], k_1 : Specific degradation rate constants of VOC [m^3/kgMLSS.h], k_2 : Rate constant [l/mg], V_g : Volume of gel beads[m^3], V_l : Deionized water volume [m^3], and V_r : Total reactor volume($V_r = V_g + V_l$)[m^3]

Table 2
Design equation for removal in packed column

$$\frac{p_{out}}{p_{in}} = \exp\left(\frac{-\eta k_1 XRTV_{Gel}}{F_G He} + k_2\left(\frac{p_{in} - p_{out}}{He}\right)\right) \quad (3)$$

For low inlet VOC concentration with first-order degradation rate equation

$$\frac{P_{out}}{p_{in}} = \exp\left(\frac{-\eta k_1 XRTV_{Gel}}{F_G He}\right) \quad (3a)$$

Fractional removal

$$E = 1 - \frac{P_{out}}{p_{in}} = 1 - \frac{C_{out}}{C_{in}} \quad (4)$$

where, E : Fractional removal[-], P_{in}, P_{out}: Partial pressure of VOC at inlet and outlet [atm],C_{in} ,C_{out} : Concentration of VOC at inlet and outlet [ppm_v], T : Operating temperature[K], V_{Gel}: Volume of gel beads in packed column[m^3], R : Gas law constant [$atm. \ell /mol.K$], He : Henry's law constant[$atm. \ell /mol$], F_G : Gas flow rate[m^3/h] and η : Contact efficiency of gel beads[-]

3. EXPERIMENTAL

3.1 Measurement of basic parameters

For measuring the degradation rate constant of k_1 and k_2, a fixed quantity of the gel beads and a known dissolved VOC concentration, C_A were added to the 100ml air- tight flask. The time course of C_A was followed by gas chromatography. For Henry's law constant, the equilibrium of VOC concentrations in the gas and liquid phases, C_g and C_l were analyzed to obtain the constant, $He = (C_g/C_l) RT$.

3.2 Measurement of VOC removal

A schematic diagram of the apparatus is shown in **Figure 2**. The packed column was made of transparent resin and packed with the bead and ball mixtures with volume ratio 1:2. The gel beads were the same ones as used in the previous work(Nakao et al., 1994). The operating temperatures were the ambient ones.

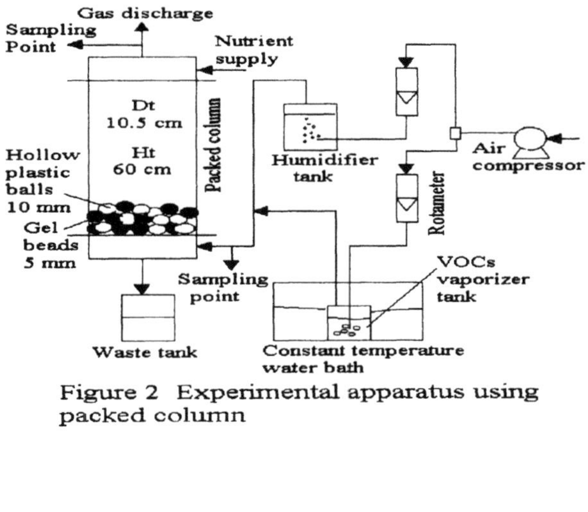

Figure 2 Experimental apparatus using packed column

The inlet and outlet VOC gas concentrations were measured to obtain the observed removal $E_{obs} = 1\text{-}(C_{out} / C_{in})$. The time course of E_{obs} was followed for one year. During operation, the gel bead activity, k_1 was measured and the calculated removal from equation(3), E_{cal} was compared with E_{obs}.

4. RESULTS AND DISCUSSION

4.1 Biodegradation kinetics

For the gel beads without acclimation, for instance, the rate constant for toluene, $k_{1,T}$ ranged from 2.22 at 20°C to 3.52 at 30°C. The constant for benzene, $k_{1,B}$ from 2.25 to 3.4.

4.2 Henry's law constant

The *He* values for toluene were from 0.17 at 20°C to 0.29 at 30°C. Those for benzene and ethylbenzene from 0.24 to 0.49 and 0.23 to 0.39.

4.3 VOC removal in packed column

Figure 3 shows the result on the toluene removal together with the k_1 values determined. The packed column was operated under the three different flowrates with the inlet concentration of 50 ppm. Removal efficiencies in the initial stage shows an unsteady-state behavior due to an acclimation of the microorganisms to toluene, and the removal appears to be stabilized after about 10 days. This reveals that the acclimation can be established in the packed column during start-up period. The observed removal agreed with the predicted ones from equations. (3a) and (4) with η equal to 0.24(Nakao et al., 1994). **Figure 4** shows the results on the ethylbenzene removal for about one year. It is seen that the observed removals decrease initially since the operation was shifted from the previous removal of toluene. Nevertheless, the observed removal shows a steady-state behavior thereafter, and suggests satisfactory performance without an inhibitory effect of the biological activity in the gel beads. The observed removals are seen to deviate from the design equation (3a). This is may be due to the fact that the inlet concentration of ethylbenzene was high enough to make high the equilibrium VOC concentration in water within the gel beads. Therefore, the design equation (3) was used to predict the removals. It is seen in **Figure 4** that the calculated removals agree with the observed ones.

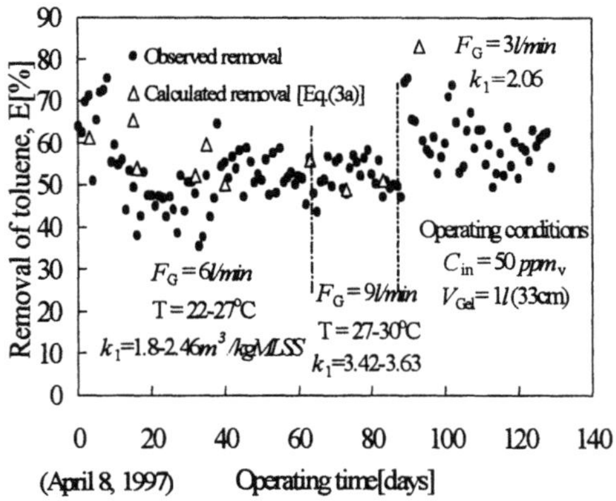

Figure 3 Time course of fractional removal of toluene

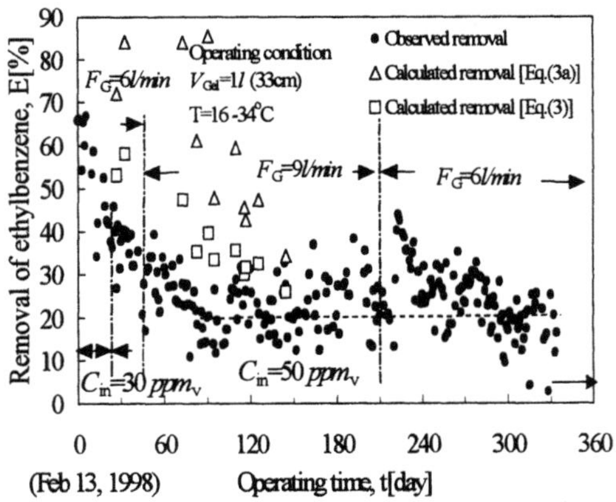

Figure 4 Time course of fractional removal of ethylbenzene

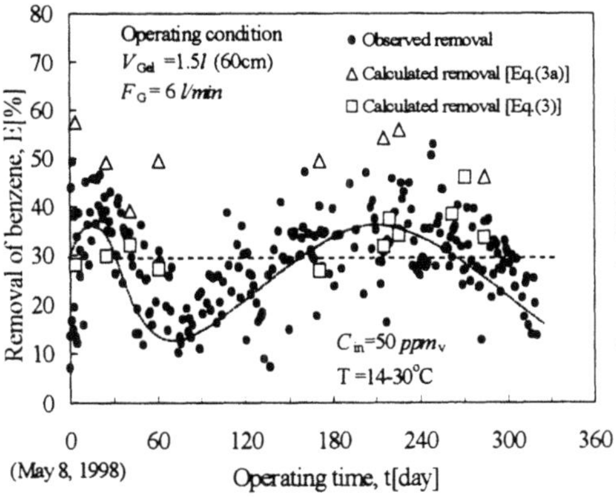

Figure 5 Time course of fractional removal of benzene

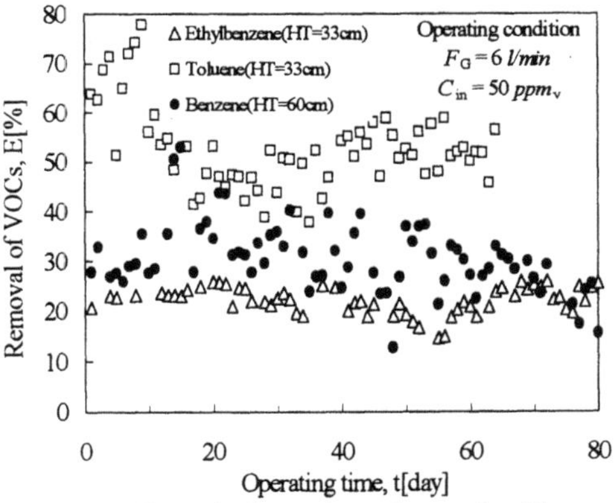

Figure 6 Comparison of removals of three VOCs during steady-state removal

The similar results were reported in the case of acetaldehyde and propionaldehyde removals (Nakao et. al., 1996). **Figure 5** shows the result on the benzene removal for about one year. It indicates that only 2 weeks are required for acclimation of benzene removal. The observed removals are seen to fluctuate during the operation. The seasonal change in the ambient temperature, i.e. the operating temperature, is thought to be the reason for the fluctuation of removals. The observed removals are seen to be in a better agreement with the removals calculated from equation (3) than those from equation (3a).

Figure 6 shows a comparison of the observed removals of toluene, ethylbenzene and benzene during steady-state removal with the identical gas flowrate and inlet concentration at almost the same temperature. It is seen that the average toluene removal is higher than the average removal of benzene and ethylbenzene. It is noted that the ethylbenzene is easier to remove than benzene, since the column height for the former removal was lower than that for the latter removal. Thus, it is suggested that E(toluene)> E(ethylbenzene)> E(benzene) at the identical operating conditions.

5. CONCLUSION

The results on the long-term removal operation indicated that the PVA-immobilized activated sludge gel beads were reproducible, stable and durable as the biocatalysts for removing the VOC. The present packed column containing the gel beads as well as hollow plastic balls was efficient for the waste gas treatment process to remove the VOC and its removal performance could be predicted by the design equations proposed.

REFERENCES

1. K. Nakao, Y. Yasuda, K .Fukunaga and M. Kimura, "Deodorization of waste gas with immobilized activated sludge, " Proc, APBioChEC 90, Kyungju, Korea, (1990) 592-595.
2. T.O. William, and F.C. Miller, "Odor control using biofilters, part II."Biocycle,**33**,(1992) 75-79.

3. K. Kirchner, U. Schlachter and H. J. Rehm, "Biological purification of exhaust air using fixed bacterial monocultures, " Appl. Microbiological Biotechnol., **31**, (1989)629 – 632.
4. S. Hartmans and J. Tramper, "Dichloro methane removal from waste gases with a trickle - bed bioreactor, " Bioprocess Eng., **6**, (1991)83-92.
5. A. P . Togna and B. R. Folsom, "Removal of styrene from air using bench scale biofilter and biotrickling filter reactors, " Proc. 85th Annu. Meeting and waste Mgmt. Assn.(1992)
6. G. A. Sorial, F. L. Smith, P. J. Smith, M. T. Suidan, A. Pandit, P. Biswas and R.C .Brenner, "Development of aerobic biofilter design criteria for treating VOCs, " Proc., 86th Annu. Meeting and Exhibtion of Air and Waste Mgmt. Assn(1993).
7. P. J. Smith, P. Biswa, M.T. Sudan and R.C. Brenner, "Treatment of volatile organic compounds in waste gases using a trickling biofilter system: A modeling approach," Proc., 86th Annu. Meeting and Exhibition of Air and Waste Mgmt. Assn.(1993)
8. K. Nakao, Y. Yasuda, H. Funahashi, K, Fukunaga and M. Azuma, "Pilot scale deodorization of waste gas with immobilized activated sludge gel beads, " Proc, 3rd Asia - Pacific Biochem. Eng. Conference, (1994)849-851.
9. M. A. Ibrahim, K. Souno, Y. Yasuda, H. Funahashi, K .Fukunaga and K. Nakao, "Toluene removal from waste gas in a column packed with immobilized activated sludge gel beads," Preprints of Okayama meeting of SCEJ, (1997)211- 212.
10. K. Nakao, H. Mizuno, Y. Yasuda, H. Funahashi and K. Fukunaga, "An analysis on deodorization in a packed column with immobilized activated sludge gel beads in case of high inlet malodorant concentration," Preprints of The 61th Annual Meeting of SCEJ, No.**2**, (1996)43.

Chapter 5

Industrial Separation Processes and Validations

Bioseparation Engineering
I. Endo, T. Nagamune, S. Katoh and T. Yonemoto (Editors)

Validation of Bioprocess Chromatography: Principles and Practices

E. K. Lee and S. J. Ahn*

Department of Chemical Engineering, Hanyang University, Ansan, Korea, and
*Genetic Engineering Team, Korean Green Cross Corp., Seoul, Korea.

ABSTRACT

Validation has become a key issue in the production of biopharmaceutics or biologics intended for therapeutic use. In addition to the validation of final product quality, manufacturing process validation is gaining wider attention. Bioprocesses are delicate and sensitive in nature, and thus careful planning and organization are required for any successful process validation. Bioprocess chromatography is probably the most widely used unit operation in biologics manufacturing. In this paper, using ion exchange as a model process, we present the principles and the practices involved in the chromatography validation.

Key words: validation, bioprocess chromatography, ion-exchange, biologics manufacturing

1. INTRODUCTION

For a successful biologics plant, the following elements should be validated: facilities, utilities, equipments (supporting and manufacturing), manufacturing processes (including cleaning), test methods, final product, and computer systems. Validation is defined as "to establish documented evidence which provides a high degree of assurance that a specific process will consistently produce a product meeting its predetermined specifications and quality characteristics, i.e., identity, strength, purity, stability, and safety"[1]. Thus, the four key elements of validation are documentation, process specificity, product specificity, and quality. The significance of process validation and its important aspects have been adequately described in the literature [2-3].

Bioprocess validation is based on the 'product-by-process' concept, i.e., the quality of a biological product is strongly affected by a manufacturing process employed to produce it [1]. Successful bioprocess validation means presenting clearly and fully the documented evidence that a specific process claims to achieve certain quality attributes of a specific product. It is viewed as a solid foundation for cGMP for quality assurance [4].

Generally, process validation consists of the sequential procedures for the validation of the equipments used for the process, i.e., IQ (installation qualification) and OQ (operational qualification), followed by the validation of the process itself, PQ (performance qualification). Process chromatography is probably the most widely used bioseparation unit operation in biologics manufacturing [5]. It is primarily used in the final purification steps for impurities removal and/or product fractionation.

2. BASIC STEPS OF PROCESS VAIDATION

There is no universally accepted procedure of process validation, but it usually consists of the following sequential steps.

(1) *Define the scopes and objectives of a process.* What a process can achieve and how it is operated must be clearly defined.

(2) *Identify the critical process parameters and their operating ranges.* The critical parameters are those that directly influence the product quality, and they must be identified carefully. And, the operating ranges of them are determined either experimentally or theoretically.

(3) *Identify probable consequences to determine the acceptable range of criteria.* Identify the parameters that can best describe the process outcome. Then, for each output parameter, estimate and/or measure it quantitatively to set a range that a successful process can achieve.

(4) *Devise a scientifically sound protocol to verify the input-output relationship.* Once the operating ranges of the input parameters and the acceptable ranges of the output consequences are determined, a simple but logical experimental protocol should be written to verify the relationship. Any analytical test procedure employed to determine process consequences should be stated or referenced.

(5) *Collect and then analyze experimental data to support the process consequences in reproducibility and in quality.* Execute the protocol and gather data to evidence that a process can reproducibly meet the predetermined quality attributes. Usually if three consecutive tests turn out successful, the validation is deemed complete.

3. PROCESS ION-EXCHANGE CHROMATOGRAPHY

The main objective of a bioprocess chromatography is to separate and remove process contaminants such as impurity proteins, endotoxins, nucleic acids, host cell proteins, process additives, etc., in manufacturing scale. Several types of process chromatographies are used in biologics manufacturing. They include but are not restricted to ion exchange, hydrophobic interaction, affinity, gel permeation, and preparative HPLC. Among these, ion exchange chromatography is probably the most widely used, and we will use it as a model chromatography to explain the principles and practices of process validation.

4. CRITICAL PARAMETERS IN ION-EXCHANGE CHROMATOGRAPHY

The performance of bioprocess chromatography is ultimately judged by recovery yield and product purity. As such, the critical parameters to be validated can be classified into two categories: process-specific and non-specific. The former includes resolution efficiency and capacity, product purity, contaminants removal efficiency, etc. The latter can be further divided into two groups: buffer-related and column-related. The buffer-related parameters can be pH, conductivity, and temperature, among others. Validation of these parameters is rather straightforward. Murphy and Seely [6] proposed an excellent protocol to validate that these parameters were maintained in the predetermined ranges particularly during hold times. The column-related parameters, which include column packing, washing, regeneration and cleaning performance, leakage from media, and others, are relatively independent of the process and/or product used. The objectives, general procedures, and acceptance criteria of validation of these parameters are discussed.

5. VALIDATION OF CRITICAL PARAMETERS

5.1. Column Packing

For ion exchange column, HETP (height equivalent to a theoretical plate) is measured usually by conductivity spiking method and its value is compared against the mean bead diameter [5]. If it is lower than approximately twice the mean diameter, a column is considered well packed. HETP needs to be measured after each packing. It is essential to use the same HETP measurement method each time and to trace its value after each cycle. It should be noted that HETP can only indicate how well packed the bed is and not the condition of the resin [7].

5.2. Column washing

The objective of washing validation is to demonstrate that a washing procedure can reduce the impurity, particularly proteineous materials, concentration to sufficiently low level. The usual procedure is to sample aliquot of final wash eluate or mock eluate buffer after the washing step and assay for the major impurity. For biologics manufacturing, the impurities are assayed as total proteins, but when the major impurities are not proteineous materials other appropriate analytical methods should be used for quantification. The usual acceptance limit is less than 1/1,000 of single dose.

5.3. Column regeneration

The objective is to confirm that after the regeneration procedure the resin maintains the majority of the original adsorption capacity. For ion exchange column, aliquot of the regenerated resin is sampled and titrated with HCl or NaOH to check the binding capacity in meq/ml. Scale-down experiments are also accepted, in which all the operational conditions are set the same as the manufacturing scale. In general, it is accepted if more than 75% of the initial capacity is maintained after each regeneration. It is important to keep the historical plot on the capacity to monitor the condition of the resin.

5.4. Column sanitization

Sometimes, the column needs to be sanitized; for instance, after several cycles, between different batches, after long period of storage, or before product changeover. The purpose is to demonstrate the given sanitization procedure routinely reduces the bioburden and the impurities to sufficiently low levels. Similar to the washing validation procedure, aliquot of the final eluate or mock eluate is sampled and assayed for the bioburden and major impurity or product residue from the previous batch. Bioburden level and total protein (or product residue) concentration are kept lower than 1 cfu/100ml and 1/1,000 of a dose, respectively.

5.5. Column lifetime

Column life is strongly dependent on each product and process, particularly the clarity of the feed material and the types of the adsorbed materials. Two approaches can be taken to determine and validate a resin's maximum lifetime. One is to make retrospective determination based on the historical plots of adsorption capacity, pressure drop, and yield/purity profiles, etc. The other is to use a miniaturized column, usually 1/1,000 of the manufacturing scale, for ex-situ determination. Since the lifetime varies widely depending on the product and process, the acceptable criteria to determine the lifetime can be also different.

5.6. Removal of endotoxin or host cell proteins

Bioprocess chromatography is widely used to eliminate key impurities such as bacterial endotoxin and host cell proteins and nucleic acids. The objective here is to show that a given chromatographic step can consistently remove those impurities down to sufficiently low levels. The purification performance may vary depending on the product and/or process, and thus universally accepted criteria are not available. This type of validation is usually performed using challenge (or spike) test, in which known amount of the target impurity is introduced into an ex-situ, miniaturized column. The mini column is operated exactly the same manner as the manufacturing scale. The eluate is assayed for the impurity to determine its removal performance. Usually, greater than 3 log reduction in endotoxin concentration is regarded satisfactory. Each regulatory authority adopts slightly different criteria for endotoxin in the final product, but it is around 5 EU (endotoxin unit) per dose. For HCP, it is less than 2% of the total protein, and for host cell DNA, less than 10 pg per dose is acceptable. Again, retrospective validation based on the historical plots can be accepted, if it shows the process reproducibly controls the target impurity level within the acceptable range.

5. CONCLUSIONS

For successful bioprocess validation, the input and output parameters of a given process are first carefully identified and the relationships between them are scientifically and systematically elucidated. Then, relevant experiments are carefully planned and executed and their results are documented, in order to provide evidence that the given process can consistently and confidently produce a product meeting predetermined quality attributes. The validation process is a collaborative teamwork between R&D, manufacturing, process development, engineering and maintenance, QC, and QA. It should be pointed out that the industries more strongly committed to validation and quality generally show more successful businesses in the long run [8].

REFERENCES

1. Guidelines on general principles of process validation, CBER, US FDA, May 1987.
2. I.R. Berry, Practical process validation of pharmaceutical materials, Drug & Cos. Ind., 139 (1986) 36-46.
3. B.T. Loftus, The regulatory basis for process validation. In: Pharmaceutical process validation. Edited by B. Loftus and R. Nash. Marcel Dekker, New York, pp. 1-9, 1984.
4. W.L. Schwemer, Validation: foundation of GMP, Pharm. Eng., 10 (1990) 44-46.
5. G.K. Sofer and L.-E. Nystrom, Process chromatography: a guide to validation, Academic Press, London, 1991.
6. R. Murphy and R. J. Seely, Validation of biotechnology bulk pharmaceuticals. In: Validation of bulk pharmaceutical chemicals. Edited by I.R. Berry and D. Harpaz. Interpharm Press, Inc., Buffalo Grove, IL, pp. 271-300, 1997.
7. G.K. Sofer and L.-E. Nystrom (eds), Process chromatography: a practical guide, Academic Press, London, 1989.
8. M. Zett, Biopharmaceutical production: the quuality audit, BioPharm '90 Proceedings (1990) pp.21-31.

Bioseparation Engineering
I. Endo, T. Nagamune, S. Katoh and T. Yonemoto (Editors)

Column Qualification in Process Ion-Exchange Chromatography

Oliver Kaltenbrunner and Peter Watler
Department of Process Engineering, Amgen Inc, Thousand Oaks, CA, USA
Shuichi Yamamoto
Department of Chemical Engineering, Yamaguchi University, Ube, Japan

To assure consistent operation of preparative chromatography columns, an integrity test, consisting of an injection of a small amount of a tracer substance, is often used to qualify the packing quality prior to use. In many cases however, the variability of the integrity test method itself can exceed the tolerance range for qualification of the column. This makes it difficult to assess the packing quality of a column prior to use in GMP manufacturing. In order to identify a test method with acceptable variability, various methods were assessed. In many instances, the tracer substance was found to interfere with the ion exchange ligand resulting in widely varying values of HETP and asymmetry factor for the same column. To reduce test variability, selection of the equilibration buffer, the tracer substance and the test velocity were optimized. A method applicable to both anion and cation exchange resins requires equilibrating the column in 0.1 M NaCl and injecting a tracer of 1 M NaCl at a flow rate of $u_o = 3000/d_p$ to $8000/d_p$.

1 INTRODUCTION

The quality and integrity of the column packing is an important factor in achieving reproducible chromatography column performance. In GMP manufacturing where the reproducibility of column performance is often monitored(1), the quality of the column packing is often tested prior to use in order to reduce process variability. One generally accepted(2, 3, 4) parameter for describing packing quality is the height equivalent to one theoretical plate (*HETP*). HETP values are commonly used to uncover gross problems with the set-up of large-scale equipment setup and column packing(5).

A second measurement for assessing column packing is the asymmetry factor (A_f). A_f is a numerical expression of the peak skewness. The peak shape can help diagnose the cause of a certain packing anomalies(6). Peak tailing can be caused by exponential wash out kinetics to the flow profile. Peak fronting is normally caused by flow channeling and can indicate a low packing density. Recently, MRI inspection of columns confirmed that under-compression of column packing can cause peak fronting while over-compression can cause peak tailing(7).

HETP and A_f can be calculated from tracer retention and dispersion data from either pulse response or frontal analysis experiments(4). It is important that the test method measures zone spreading induced by the flow pattern through the packing while minimizing contributions from all other sources. To minimize these unwanted contributions, one must: 1) standardize the mathematical calculation 2) optimize the test velocity to minimize contributions from diffusion and mass transfer resistances, and 3) minimize the interaction of the tracer molecule with the

stationary phase. Common calculation methods standardize the zone spreading measurement. Nevertheless, it should be clear that a measured value of *HETP* is not an absolute number but reflects the applied experimental and numerical method both in its absolute value and its variability(4). The test velocity can be based on well-known principles (8, 9) which account for the diffusivity of the tracer molecule and the resin particle size. Selection of the equilibration & elution buffer and the tracer molecule is the most challenging element. In GMP manufacturing, the tracer has to be nontoxic, cost-effective and easily monitored. NaCl fulfills all necessary requirements with the exception that it interacts with ion-exchange resins. To minimize the exchange of sodium and chloride ions with the resin, the equilibrium state of the ion-exchange resin must be controlled. To minimize ion-exchange reactions between the mobile phase and resin the equilibration buffer and tracer should consist of the same co- and counter-ions(10).
An example of a very unfavorable anion-exchanger system is an acetate equilibration buffer combined with a NaCl tracer. In the equilibrated state, negatively charged acetate ions form the counter-ions with the ligand. As the NaCL tracer moves through the column, the negatively charged chloride ions replace the acetate ions which have a lower affinity for the ligand. At the tail of the tracer band, the exchange is reversed, as the original acetate ion equilibrium is restored. In this paper the adverse effects of improper selection of buffer and tracer are investigated. A method generally applicable for integrity testing of any ion-exchange resin is proposed. Additionally, for this proposed method, a simplified guideline for selecting the appropriate test flow rate is given. Furthermore, possible pressure restrictions for the optimal test flow rate are illustrated.

2. THEORY

The term used to measure dispersion in a chromatographic column is the height equivalent of a theoretical plate (*HETP*) and can be defined as,

$$HETP = \frac{L \cdot M_2^{\,2}}{(M_1')^2} \qquad \text{Equation 1}$$

The most common approximation assumes a normal distribution used is based on the retention time of the peak maximum (t_c) and the peak width at half height (w_h)

$$HETP = \frac{L}{5.545} \cdot \frac{w_h^{\,2}}{t_c^{\,2}} \qquad \text{Equation 2}$$

Band broadening as a function of flow velocity (u) can be described by the Knox equation(11) as,

$$h = a \cdot ReSc^{0.33} + \frac{b}{ReSc} + c \cdot ReSc \qquad \text{Equation 3}$$

where the reduced plate height is the plate height relative to the particle diameter (d_p)

$$h = \frac{HETP}{d_p} \qquad \text{Equation 7}$$

and the group *ReSc* is the dimensionless velocity. The terms a, b, and c of Equation 3 represent the band broadening contributions from axial dispersion, diffusion, and mass transfer limitations, respectively. The description of the deviation of the response profile from its ideal symmetrical shape can be simply described by the asymmetry factor (A_f) which is the ratio of half widths of the peak at 10 % peak height. A symmetrical peak gives $A_f = 1$. $A_f < 1$ indicates fronting peaks and $A_f > 1$ indicates tailing peaks.

3. MATERIALS AND METHODS

Columns: ID = 2.6cm, 10 cm < L < 20 cm (Pharmacia, Upsala, Sweden).
Chromatography Workstation: BioSys2000 (Beckman Instruments, Inc., Fullerton, CA, USA)
Resins: Q Sepharose FF, SP Sepharose FF, SP Sepharose HP, SP Sepharose BB, CM Sepharose FF (all Pharmacia, Upsala, Sweden).
Equilibration solutions: water, 0.05/0.1/0.2/0.3/0.4/0.5M NaCl, 0.1M NaOH, 0.1M HCl, 0.05M Tris/HCl.
Tracer solutions: 1% Acetone in water, 1M NaCl, 0.2M NaOH,
Data Analysis: UV_{280}, conductivity, and pH data were collected digitally at 2 Hz. For injections of NaCl and NaOH the conductivity trace and for injection of Acetone the UV trace were analyzed. From the response profile retention time at peak maximum (t_c), width at half height (w_h) and the half widths at 10% peak height were calculated from the signal. HETP was calculated according to Equation 2.

4. RESULTS AND DISCUSSION

4.1 *Selection and characterization of Buffer/Tracer systems*

The correct selection of equilibration buffer and tracer sample is of primary importance. Figure 1 shows the response profiles obtained from different combinations of equilibration buffer and tracer. As a reference, 1% Acetone was injected as the tracer and the UV response profile analyzed. All the profiles in Figure 1 show distortions of the either the UV or the pH profile. This is a result of ion-exchange interactions of the tracer band with the ligand. The characteristic double wave in the pH profile, visible in five of the seven profiles, indicates capture of hydronium or hydroxide ions with approaching tracer zone and release of these ions as the tracer zone departs or vice versa. The profile of injecting NaOH after equilibration with NaOH does not show any distortion of the pH trace, but the dip in the UV trace at the onset of the peak indicates possible interactions with the resin.

The system with NaCl in both the equilibration buffer and the tracer shows only a minor change in pH and no significant deformation of the conductivity profile. It is a good system since there no change in co- and counter-ions between the buffer and the tracer. All subsequent tests were performed using this system at various NaCl concentrations. To define the appropriate NaCl equilibrium buffer concentration, injections of 1M NaCl were made in anion and cation exchange columns equilibrated with buffers of varying NaCl concentrations. Results for SP Sepharose FF are also shown in Figure 1. As expected, *HETP* increases with increasing flow velocity while A_f decreases slightly with increasing velocity. This is true only for conditions, where NaCl was added to the equilibration buffer. When the column is equilibrated in water, the peaks showed fronting and completely different peak shapes, indicating an overwhelming interaction of the tracer ions with the ligand. This behavior was seen with both anion and cation exchange resins.

There is also a significant reduction in retention volume (0.2-0.3 CV) when NaCl is present in the equilibration buffer. When NaCl is abscent in the buffer, $V_R > 1$ CV, when more than 0.1M NaCl is present, $V_R \approx 0.9$ CV which was the same retention volume as acetone.

Investigating the effect of NaCl concentration in the equilibration buffer verifies the dramatic change in result from when no salt is present and when even a small amount of salt is present in the equilibration buffer. No significant change in the test results was seen when the NaCl concentration was increased from 0.1 to 0.5 M in the equilibration solution. To enhance the data analysis, i.e. have a higher signal to noise ratio in the conductivity trace, the lowest possible equilibrium concentration of NaCl should be used. A concentration of 0.1 to 0.2 M NaCl in the equilibrium buffer and a 1M NaCl tracer is the preferred system.

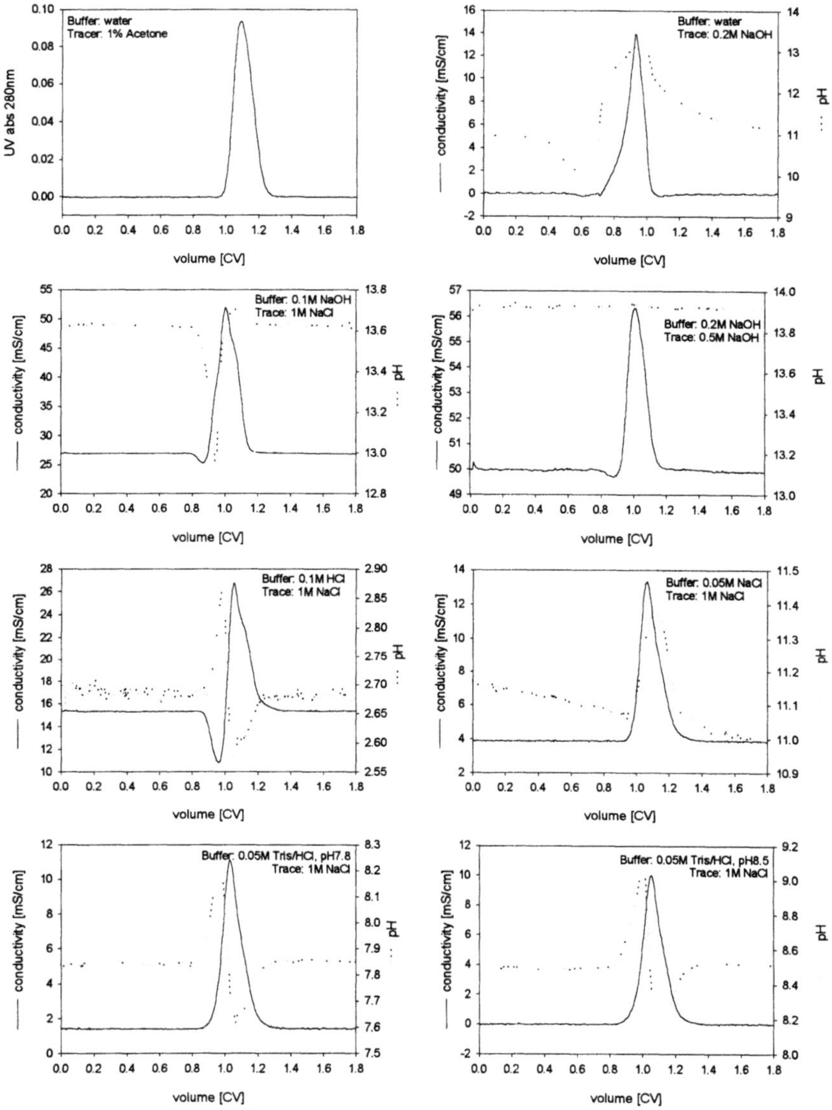

Figure 1: Comparison of column qualification methods on Q Sepharose FF. i.d.: 2.6cm; L: 15 cm; u_0 = 60 cm/h. The volume axis is shown in column volumes [CV].

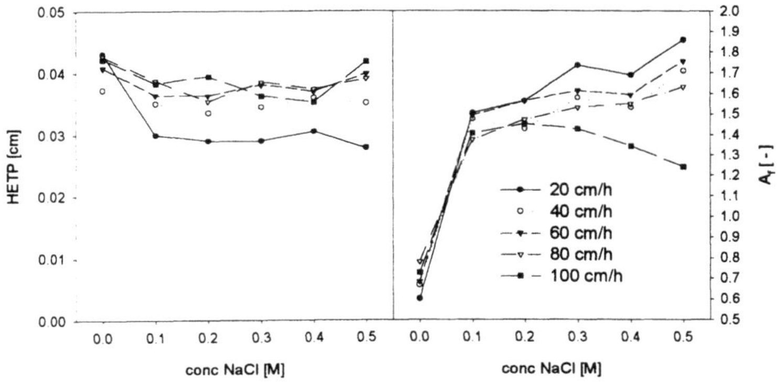

Figure 2: HETP and A_f of SP Sepharose FF for various flow velocities and equilibrium NaCl concentrations. i.d.: 2.6cm, L: 15cm, Tracer: 2% column volume of 1M NaCl

4.2Selection of test velocity

Comparison of the effects of test velocity on HETP should be independent of particle size. Using Equation 3 The dimensionless flow velocity can be expressed as

$$Re \cdot Sc = \frac{d_p \cdot \varepsilon \cdot u}{\nu} \cdot \frac{\nu}{D_m} = \frac{u_0 \cdot d_p}{D_m} \qquad \text{Equation 4}$$

and *HETP* data can be made dimensionless by eqn. 4. The optimal flow velocity for the test is when the ratio between the *a* term and the total *h* is a maximum. This ensures the greatest contribution is from the packing effects and minimizes the contribution from diffusion and mass transfer limitations. Figure 4 shows a plot of the ratio *a/h* from Equation 3 versus the flow velocity *ReSc*. Testing a column at a linear velocity of *ReSc* < 5 cannot give reliable information on the quality of the column packing. *HETP* values in this range primarily reflect diffusion of tracer ions and is strongly influenced by temperature and flow rate variations. Conversely, the higher the linear flow velocity, the greater is the contribution from mass transfer limitations (c–term) and the higher the back pressure. Hence, the best operating range is at linear velocities of 5 < *ReSc* < 15 which ensures a high contribution of the *a* term.

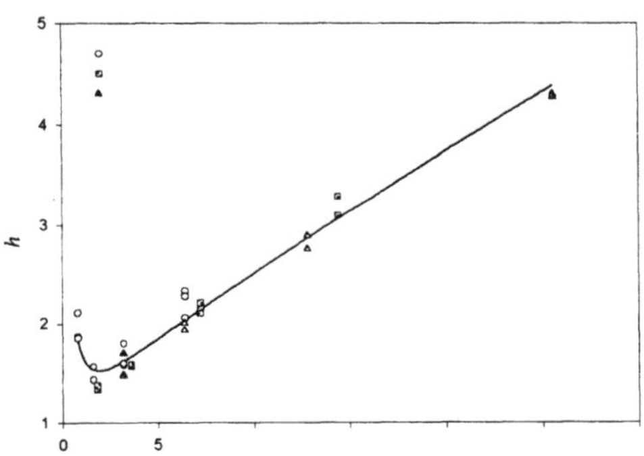

Figure 3: Column integrity test results for different particle size resins as a function of flow velocity. The data are fitted using Equation 3.

For the proposed test method, with $D_{NaCl,water} = 1.5\ 10^{-5}\ cm^2s^{-1}$, the superficial velocity (u_0) can be estimated as

$$\frac{3000}{d_p} < u_0 < \frac{8000}{d_p} \qquad \text{Equation 5}$$

where u_0 is in cm/h when d_p is in μm.

For small particles, the optimum test region is restricted by back pressure constraints. Applying the Kozeny Carman equation (12)

$$u_0 = \frac{\varepsilon^3}{270 \cdot (1-\varepsilon)^2} \cdot \frac{d_p^{\,2}}{\mu} \cdot \frac{\Delta p}{L} \qquad (6)$$

allows for an estimate of the maximum column length (L_{max}) for a given d_p and a back pressure (Δp_{max}) limitation. When equations 6 and 4 are combined, L_{max} to allow $ReSc = 5$ at Δp_{max} is

$$L_{\max} = \frac{\Delta p_{\max}}{\mu} \cdot \frac{d_p^{\,3}}{1350 \cdot D_m} \cdot \frac{\varepsilon^3}{(1-\varepsilon)^2} \qquad (7)$$

For the test method with injections of 1M NaCl in a 0.1 to 0.2M NaCl equilibration buffer at 20°C ($\mu_{0.2M\ NaCl,\ 20°C} = 1.024$ cP) and assuming $\varepsilon = 0.4$ and $D_m = 1.5\ 10^{-5}$ cm^{2}s^{-1}, the equation simplifies to

$$L_{\max} = \frac{\Delta p_{\max} \cdot d_p^{\,3}}{1.25 \cdot 10^6} \qquad (8)$$

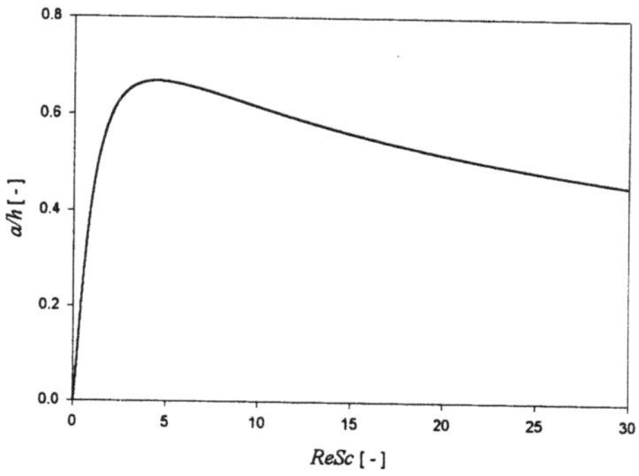

Figure 4: Contribution of the *a* term of Equation 3 to the total broadening *h*.

5. Conclusion

To increase consistency in column integrity testing of ion-exchange columns, the column should be equilibrated with the same counter ion that is used for the test. This avoids excessive ion exchange effects between the resin ligand and the test tracer. As a method generally applicable to both anion and cation exchange resins, a system of a 1M NaCl tracer in a column equilibrated with 0.1M NaCl is recommended. The optimum flow rate range for column testing is from $ReSc = 5$ to 15. For the test system with NaCl, this translates to u [cm/h] = $3000/d_p$ to $8000/d_p$. (d_p in μm). Generally, this means, the larger the particle, the lower the test flow rate. When operated in this manner, *HETP* should be smaller than 5 d_p. Pressure limitations can arise for smaller beads. If the test cannot be executed at the appropriate flow rate, higher values for *HETP* are to be expected.

References

1. G. K. Sofer, L.-E. Nyström, *Process Chromatography. A Guide to Validation* (Academic Press, London, 1991).
2. G. Guiochon, *J. Chrom.* **185**, 3-26 (1979).
3. L. R. Snyder, *J. Chrom. Sci.* **10**, 369-379 (1972).
4. H. P. Lettner, O. Kaltenbrunner, A. Jungbauer, *J. Chrom. Sci.* **33**, 451-457 (1995).
5. A. Jungbauer, E. Boschetti, *J. Chrom. B* **662**, 143-179 (1994).
6. N. S. Mitchell, L. Hagel, E. J. Fernandez, *J. Chromatogr., A* **779**, 73-89 (1997).
7. M. L. Dickson, P. Leijon, L. Hagel, E. J. Fernandez, Revealing Packing Heterogeneities and Their Causes Using Magnetic Resonance Imaging, Prep'98, Washington, DC, USA (1998).
8. J. J. van Deemter, F. J. Zuiderweg, A. Klinkenberg, *Chemical Engineering Science* **5**, 271-289 (1956).
9. J. C. Giddings, *Dynamics in Chromatography. Part I. Principles and Theory*, Chromatographic Science Series (Marcel Dekker, Inc., New York, 1965), vol. 1.
10. F. Helfferich, *Ion exchange* (McGraw-Hill, New York, 1962).
11. J. H. Knox, in *Advances in Chromatography* P. R. Brown, E. Grushka, Eds. (Marcell Dekker, New York, 1998), vol. 38, pp. 1-49.
12. J. C. Giddings, *Unified Separation Science* (John Wiley & Sons, New York, 1991).

Bioseparation Engineering
I. Endo, T. Nagamune, S. Katoh and T. Yonemoto (Editors)

Characterization of Phage Encoded Lysis Proteins and Its Applications for Cell Disruption

Yasunori TANJI, Katsutoshi HORI, Shinjiro YAMAMOTO, and Hajime UNNO

Department of Bioengineering , Faculty of Biotechnology, Tokyo Institute of Technology
4259 Nagatsuda-cho, Midori-ku, Yokohama 226-8501, Japan

Using a lysis assay a gene, lys1521, was isolated from a bacteriophage specific to *Bacillus amyloliquefaciens* and its nucleotide sequence revealed one open reading frame of 375 bp. Overexpression of the cloned gene yielded a 13 kDa protein corresponding to the predicted gene product. The lytic profile obtained in an in vivo expression assay showed that lys1521 had cell wall hydrolysis activity.

Two lysis genes, gene-t and gene-e, encoded in bacteriophage T4 were also cloned into pET26b(+) vector. Those two lysis genes were expressed in BL21(DE3) *E. coli* cells. Immediate cell disruption was observed when the gene-t was expressed in the logarithmic growth phase. In an attempt to develop a self disruptive *E. coli* expression system, gene-e, was cloned into expression vector pACYC184. The β-glucuronidase (GUS) was selected as a model target protein. The production of gp-e with the N-terninal fusion of signal peptide of pelB did not cause immediate cell lysis. Resuspension of this gene product-e (gp-e) producing BL21(DE3)pLysS cell pellet with pure water caused cell lysis followed by GUS release to the medium. Almost 50 % activity of the GUS was identified in the resuspended supernatant.

1. INTRODUCTION

Many useful proteins such as hormones, enzymes and vaccines produced by using *E. coli* protein expression systems. However, *E. coli* is not a perfect host. The major problems results from the fact that *E. coli* does not normally secrete proteins into the extracellular medium. Therefore, the first step of the protein recovery procedure is always the disruption of the cells. To overcome this problems, we found out a lysis enzymes from phages specific to target host cell for the application of cell disruption. Phage-encoded lysins or endolysins specific to *E. coli*, *Bacillus* are elucidated.

In this study a novel method for cell disintegration is also proposed. This method is based on the expression of a cloned gene of T4 phage lysis proteins resulting in gentle lysis of *E. coli*. T4 phage is one of the *E. coli* specific virulent phage. Lysis by T4 phage requires the action of two gene products, E and T (1). Gene-e encodes for a lysozyme which degrades rigid cell wall material (peptidoglycan) of *E. coli*. Gene-t is considered to degrade or alter the cytoplasmic membrane thus allowing gene-e product (gp-e) to reach the periplasm and gain access to the peptidoglycan layer. In an attempt to develop a self disruptive *E. coli* expression system, T4 phage encoded two lysis genes were cloned into expression vectors. Lysis profiles of *E. coli* harboring two lysis genes were compared. Efficiencies of cell disruption evaluated by GUS release to the medium were also discussed.

2. EXPERIMENTAL

2.1. Isolation of the *Bacillus* phage and cloning a lysis gene

B. amyloliquefaciens (IAM 1521) was used as a propagating host. Activated sludge was obtained from a local sewage treatment plant. Bacteriophages were isolated using the plaque assay. Phage DNA was digested with the restriction endonuclease *Hin*dIII. Fragments ranging from 1000-3000 bp were isolated and ligated into a linearized pUC118 expression vector. *Escherichia coli* JM109 was used as a host cell in recombinant manipulations. All *E. coli* cells carrying the recombinant plasmid were plated onto standard Luria-Bertani (LB) agar medium supplemented with 100 mg ml^{-1} ampicillin. The lysis assay was performed as described (2). Positive colonies were isolated from the original plate and their DNA was analyzed by endonuclease digestion.

2.2. Controlled Expression of T4 Phage Encoded Lysis Genes

To make T4 phage lysis gene expression plasmids, DNA fragments of lysis gene were obtained by the polymerase chain reaction (PCR). To obtain the plasmids of pET26b/E and pET26b/T, *Bam*HI-*Hin*dIII fragments of two PCR products were inserted into the *Bam*HI-*Hin*dIII site of pET26b(+)(Novagene).

2.3. Dual expression system

To make GUS and T4 phage lysis gene expression plasmids, DNA fragments of these two genes were obtained by the PCR and integrated into two plasmids, pKF4 and pACYC184, respectively. The resultant plasmids were pKF4GUS and pACYC184E. Expression of the GUS gene encoded in pKF4GUS vector is under control of trp promoter: expression is induced by incubating transformant *E. coli* cells in tryptophan free medium. On the other hand, expression of the T4 phage lysis gene encoded in pACYC184E vector is under control of bacteriophage T7 transcription and translation signals: expression is induced by providing a source of T7 RNA polymerase in the host cell of *E. coli* BL21(DE3). Plasmid pACYC184E carries a T7 lysozyme gene which is a natural inhibitor of T7 RNA polymerase, and thus reduces its ability to transcribe target genes in uninduced cells. Vector, pACYC184E, allows fusion of N-terminal signal sequence of pelB outer membrane protein of *E. coli* for periplasmic localization of the gene-e product.

3. RESULTS and DISCUSSION

3.1. Lytic action of *Bacillus amyloliquefaciens* phage lysis protein

A bacteriophage specific to *Bacillus amyloliquefaciens*, a gram-positive bacterium, was isolated from a local sewage treatment center. Using a lysis assay, a gene, *lys1521*, was isolated and its nucleotide sequence revealed one open reading frame of 375 bp. The expression of lys1521 encoded in *E. coli* JM109 exhibited no effect on the growth rate (Fig. 1

A). However, addition of 2% chloroform, 2 h after induction of lys1521 and the control cultures caused massive lysis of JM109 cells harboring pUClys1521 (Fig. 1B). Chloroform can functionally replace holin protein by permeabilizing the inner membrane, thereby making it possible for the lys1521 protein to reach the peptidoglycan layer and cause hydrolysis. To characterize the production and the transfer of the protein from the cytosol to the peptidoglycan, lys1521 was cloned into the pET26(b) vector. Induction and subsequent incubation effectively decreased growth in the culture harboring pET26lys1521 (Fig. 1C). Moreover, massive lysis occurred by the addition of 2% chloroform (Fig. 1D). Therefore, the lys1521 protein could be localized in the cytosol and once it was presented to the peptidoglycan served to bring about functional hydrolysis activity in *E. coli*..

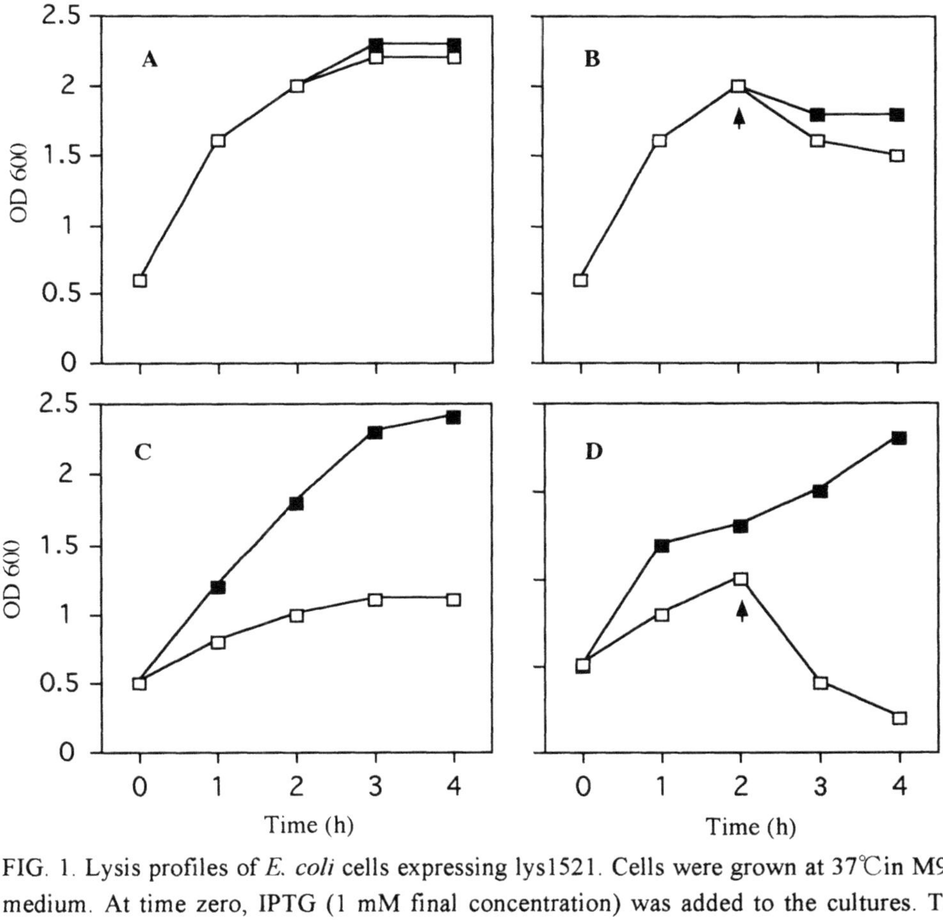

FIG. 1. Lysis profiles of *E. coli* cells expressing lys1521. Cells were grown at 37℃in M9G medium. At time zero, IPTG (1 mM final concentration) was added to the cultures. The strains used were: JM109 with pUClys1521 (A, B) and JM109(DE3) with pET26lys1521 (C, D). At the time indicated by the arrow, 2% chloroform was added to the cultures to induce endolysin leakage thus causing cell lysis. Symbols: □, IPTG-induced cultures; ■, non-induced cultures.

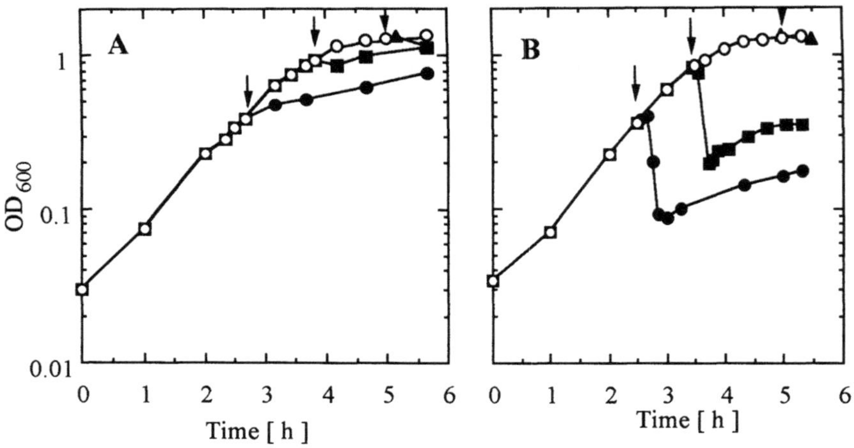

Fig. 2. Profiles of *E. coli* cell lysis by expressing pET26b encoded T4-phage lysis genes, gene-e (A) and gene-t (B). At an OD_{600} of 0.4 (●), 0.8 (■), and 1.3 (▲), IPTG was added to give a final concentration of 1 mM; ○: control without IPTG.

3.2. Lysis profiles of *E. coli* cells by expressing T4-phage lysis genes

T4 phage lysis genes were cloned into pET vector plasmid. Expression of the target gene encoded in this vector is under control of bacteriophage T7 transcription and translation signals: expression is induced by providing a source of T7 RNA polymerase in the host cell of *E. coli* BL21(DE3)pLysS. BL21(DE3)pLysS contains a plasmid carrying a T7 lysozyme gene which is a natural inhibitor of T7 RNA polymerase, and thus reduces its ability to transcribe target genes in uninduced cells. Vector, pET26b(+), allows fusion to N-terminal signal sequence of pelB outer membrane protein of *E. coli* for periplasmic localization of the produced protein. However the OD_{600} of BL21(DE3)pLysS harboring pET26b/E cell culture after induction at different time points was not reduced (Fig. 2A), immediate stop of increase of visible cell number was observed through microscopy. In addition, phase contrast microscopy revealed morphological change of the pET26b/E expressing cells. The shape of the cells producing gp-e changed to swollen ellipsoid. Those microscopic observations envisaged that the conveyance of the gp-e produced in the cytosol to the periplasmic space facilitated enzymatic degradation of the murein layer which is responsible for maintaining the cell strength.

The production of gp-t encoded in pET26b/T, gave drastic change to the *E. coli* BL21 (DE3)pLysS cells. The bacterial culture showed a rapid turbidity drop when IPTG was added in the middle or late logarithmic growth phase (Fig.2B). Microscopic observation showed that the induction of the gp-t in the stationary phase (OD_{600} = 1.3) led part of the cell lysis. Since the produced *E. coli* pieces by the gp-t lysis action at the stationary phase were relatively large, turbidity drop of the culture was not large. It was supposed that production of gp-t in the BL21(DE3)pLysS cells allowed T7 lysozyme, encoded in the pLysS plasmid, to reach the periplasm and lead immediate cell lysis.

3.3. Dual expression system

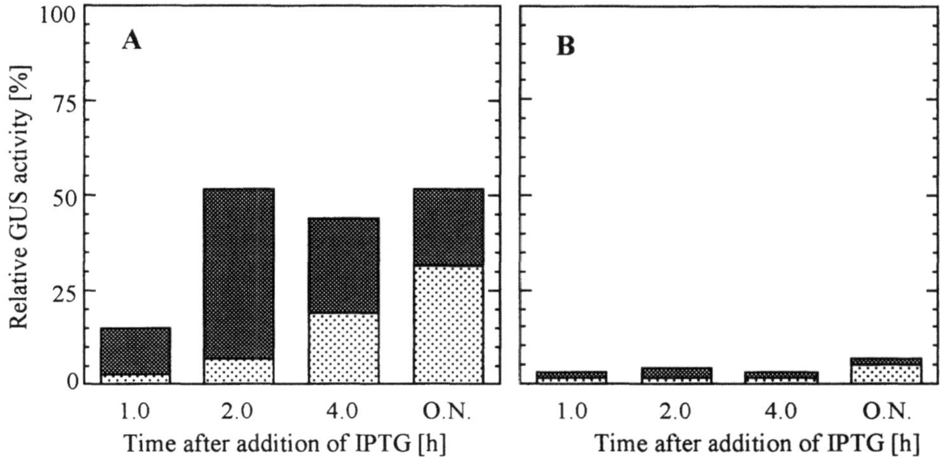

Fig. 3. Expression of T4 phage lysis gene in *E. coli* BL21(DE3)pLysS cells and secretion of GUS. Cells harboring pACYC184E were incubated until OD600 reached 0.5. A: IPTG was added to give a final concentration of 1 mM at time 0. B: control without IPTG. ■: time course of relative activity of GUS in the supernatant, ⊡ : time course of relative activity of GUS in the supernatant of the resuspended liquid. O.N.: Over night.

To identify the gp-e action for the protein release to the outside of the cell, GUS production in the BL21(DE3)pLysS and release to the medium were analyzed. When the gp-e production was induced at OD_{600} = 0.5, only a small activity of the GUS was observed in the culture (Fig. 3), indicating production of gp-e followed by secretion to the periplasmic space did not cause immediate cell lysis. On the other hand, when the pellet of gp-e producing cells were collected by centrifugation and resuspended with pure water, about 50% of the GUS activity was observed in the supernatant of the resuspended liquid after 2 hours induction period. In contrast, almost no activity was observed in the resuspended liquid when the cells without IPTG addition were treated with the same manner, indicating that simply the osmotic shock did not lead cell to disrupt (Fig. 3 B). Those facts convinced that the gp-e production in the cytosol followed by translocation to the periplasmic space by signal peptide weaken the cell strength which did not lead the prompt cell disruption. But osmotic shock in combination with physical stress such as osmotic shock facilitated lysis of those cells.

REFERENCES

1. R. Young, Microbiological review, 56, (1992) 430-481.
2. M. Loessner, Wendlinger, G., and Schere, S. Microbiol., 16, (1995) 1231-1241.

Bioseparation Engineering
I. Endo, T. Nagamune, S. Katoh and T. Yonemoto (Editors)

Recovery of poly-β-hydroxybutyrate from recombinant *Escherichia coli* by a combined biologi-chemical method*

J. Yin[a], Y. Xu[b], H.-M. Yu[a], P.-J. Zhou[b] and Z.-Y. Shen[a+]

[a]Department of Chemical Engineering, Tsinghua University, Beijing 100084, China

[b]Institute of Microbiology, Academia Sinica, Beijing 100080, China

A combined biologi-chemical method using cloned lambda phage lysis genes for cell disruption and poly- β -hydroxybutyrate (PHB) separation was proposed and studied. The new plasmid named pTU14 with the lysis genes S(-)RRz of λ DNA (cI857 sam7) and the PHB biosynthesis genes *phb*CAB from *Alcaligenes eutrophus* H16 was constructed and transformed into *E. coli* JM109. The recombinant *E. coli* JM109(pTU14) cells in stationary phase could be induced to lyse by buffer A (2 mmol/L EDTA in 50 mmol/L Tris buffer, pH=8) treatment and PHB granules were released completely with the digestion and dissolving of most of the non-PHB materials. PHB content could be increased from 66.07% to 90.53%. Further purification of PHB could be realized by surfactant SDS treatment, and PHB granules of 99.66% purity were obtained.

1. INTRODUCTION

Separation of poly-β-hydroxybutyrate (PHB) from its production organism is challenged by it being an intracellular material accumulated as granules. A number of previously reported recovery processes involved extraction of PHB with organic solvents[1], digestion of non-PHB materials by sodium hypochlorite[2] or enzymes[3].

Prior to us, the use of cloned bacteriophage lysis gene for PHB release from *E. coli* was proposed. Lubitz[4] transformed the lysis gene of phage ΦX174 into a PHB producing *E. coli* and put the lysis gene under the control of a temperature-regulated promoter. Upon induction of the lysis gene by heating to 42℃, cell lysis was completed within 10 minutes, but efficient lysis could only be obtained during exponential growth phase prior to the time when maximal

*This work was supported by grants of "National Natural Science Foundation" (No.: 29476242) and "Ninth Five Year Plan Research Program" (No.: 96-c03-03-02) of China. +Corresponding author.

PHB accumulation was obtained. Dennis[5] developed another system in which host cells had two plasmids: one containing the T7 bacteriophage lysozyme gene and the another containing the PHB biosynthesis genes. At the end of the PHB accumulation phase of the culture, cells could be lysed and PHB granules released efficiently after being pelleted and resuspended in 50 mmol/L Tris/2 mmol/L EDTA and a final concentration of 0.1% Triton X-100.

The cell lysis of lambda (λ)-infected *E. coli* is proved to be made by three λ phage lysis gene products[6]: S, R, and Rz. The R gene product, a transglycosylase, and the Rz gene product, which may be an endopeptidase, are responsible for degrading the host cell wall. The S gene product appears to be responsible for altering the cytoplasmic membrane, thus allowing R and Rz gene products to reach the cell wall. The present paper describes an improved biologi-chemical process whereby a one-plasmid system, which contains the lambda (λ) phage lysis genes S(-)RRz and the biosynthesis genes of PHB, is developed.

2. MATERIALS AND METHODS

2.1. Bacterial strains, plasmids and enzymes

E. coli JM109 was obtained from Promega. Plasmid pUC18 and λ DNA (cI857 sam7) were obtained from Sino-American Biotechnology Company. Plasmid pTZ18u-PHB which harboured PHB-biosynthesis genes *phb*CAB from *Alcaligenes eutrophus* H16 was provided generously by Professor D. Dennis in James Madison University. All restriction endonucleases and T4 DNA ligase were obtained from Boehringer Mannheim.

2.2. Isolation, analysis and transfer of DNA

DNA was isolated, digested or analyzed by standard procedures described by Sambrook *et al*[7]. Transformation of *E. coli* JM109 was done by the $CaCl_2$ procedure, as described by Sambrook *et al*[7].

2.3. Growth of bacteria

Cells were pre-grown in 5 ml medium C (Tryptone 5.0 g/L, yeast extract 1.0 g/L, beef extract 3g/L, NaCl 5.0 g/L) with 100 μg ampicillin/ml at 220 r/min and 37℃ for 10 hours, the culture was inoculated at 1:50 into 50 ml medium C with 20 g glucose/L and 100 μg ampicillin/ml in a 250 ml flask and incubated as before. Plasmid pTZ18u-PHB was used as a control of pTU14.

2.4. Procedures of cells inducible lysis

Cells at stationary phase in 50 ml medium were harvested by centrifugation at 6000 ×g for 10 min, resuspended in 25 ml buffer A (2 mmol/L EDTA in 50 mmol/L Tris buffer, pH=8) and shaken at 220 r/min and 37℃ for 20 minutes. Buffer B (50 mmol/L Tris, pH=8) was used as a control of buffer A.

2.5. Purification of PHB

Cells which were lysed by 25ml buffer A treatment were pelleted by centrifugation at 6000 ×g for 10 min, and the pellet was treated by 25ml surfactant SDS solution at 37℃ for 10 min.

The resulting concentrated PHB granules were washed once with deionized water and then dried in a 45℃ oven.

2.6. Assay of PHB contained in dry cell

Purity of recovered PHB was assayed using the method presented by Ward *et al*[8].

3. RESULTS AND DISCUSSIONS

3.1. Construction of plasmids

As shown in figure 1. Firstly, the lysis genes S(-)RRz of λ DNA (cI857 sam7) which is defective in S gene were isolated on a 1466bp *Eco*R I-*Cla* I fragment and then inserted into plasmid pUC18 which had been subjected to an *Eco*R I and *Acc* I digestion. A new plasmid named pUC18-S(-)RRz was constructed for the amplification of lysis genes S(-)RRz.

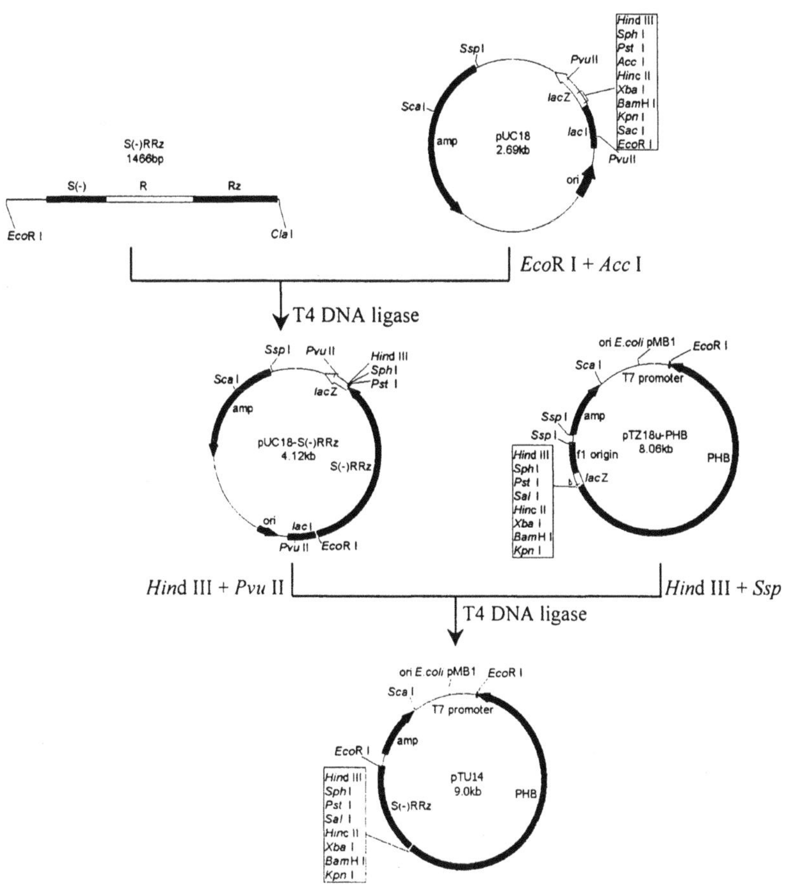

Figure 1. Construction of plasmid pTU14.

Secondly, the biosynthesis genes of PHB in plasmid pTZ18u-PHB were isolated on an *Ssp* I-*Hind* III fragment and the lysis genes S(-)RRz in plasmid pUC18-S(-)RRz were isolated on an *Hind* III-*Pvu* II fragment. These two DNA fragments were ligated by T4 DNA ligase, and a new plasmid named pTU14 was obtained, as shown in figure 1. This plasmid pTU14 which not only contains the biosynthesis genes of PHB but also contains the lysis genes S(-)RRz of λ DNA (cI857 sam7) was transformed into *E. coli* JM109 for the expressions of genes *phb*CAB and S(-)RRZ.

3.2. Inducible lysis of *E. coli* JM109(pTU14) by buffer A treatment

Because S gene is defective and can not be expressed, recombinant *E. coli* JM109 (pTU14) can grow normally with the accumalation of R and Rz gene products and PHB of *phb*CAB gene product in cells. In order to lyse the cell wall, buffer A is used to mimic the action of the S gene product to disrupt the cytoplasmic membrane. The comparison of *E. coli* JM109(pTU14) treated by buffer A and buffer B, and *E. coli* JM109(pTZ18u-PHB) treated by buffer A is shown in figure 2.

In figure 2, the decrease of cells absorbance at 600 nm for *E. coli* JM109(pTU14) resuspended in buffer A is much bigger than that for *E. coli* JM109(pTU14) resuspended in buffer B and *E. coli* JM109(pTZ18u-PHB) resuspended in buffer A. Meanwhile, observed by scanning electron-microscope, cells of *E. coli* JM109(pTU14) resuspended in buffer B and cells of *E. coli* JM109(pTZ18u-PHB) resuspended in buffer A still keep their entire shape and PHB granules remain in cells. On the contrary, cells of *E. coli* JM109(pTU14) resuspended in buffer A are lysed and the PHB granules are released completely, as shown in figure 3.

The lysis of *E. coli* JM109(pTU14) indicates that the biologi-chemical method using cloned lambda phage lysis genes S(-)RRz to disrupt cell wall for the recovery of PHB is practicable. The above results also demonstrate that compared with *E. coli* JM109(pTZ18u-PHB), only *E. coli* JM109(pTU14) can accumulate R and Rz gene products in cells. The role of R and Rz gene products is to degrade the cell wall, while the role of buffer A is the same as that of gene S product to alter the permeability of cytoplasmic membrane. In buffer A, it is not Tris but EDTA plays a major role in altering cytoplasmic membrane.

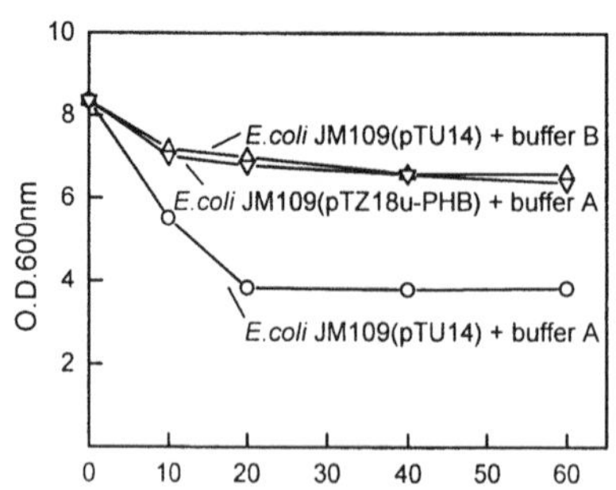

Figure 2. Comparison of cell lysis of *E. coli* JM109(pTU14) and *E. coli* JM109(pTZ18u-PHB) after buffer A and buffer B treatment.

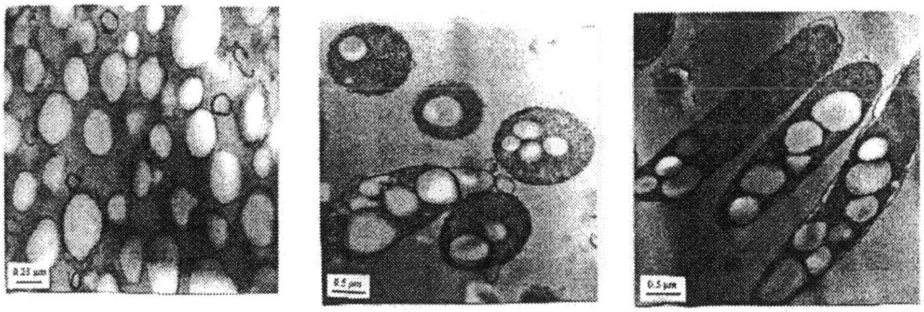

(A) (B) (C)

Figure 3. Cells of *E. coli* JM109(pTU14) and *E. coli* JM109(pTZ18u-PHB) observed by scanning electron-microscope after buffer A or buffer B treatment.
(A) *E. coli* JM109(pTU14) + buffer A (B) *E. coli* JM109(pTU14) + buffer B
(C) *E. coli* JM109(pTZ18u-PHB) + buffer A

3.3. Purification of PHB by surfactant SDS treatment

When lysed by buffer A treatment, more than half of the non-PHB materials in *E. coli* JM109(pTU14) are degraded and dissolved in the solution and the PHB content is increased too. As shown in table 1, the higher PHB content of initial cells is, the higher PHB content of cells after buffer A treatment is, and the ratio of non-PHB materials extracted is increased too. When the cells containing 66.07% PHB were treated with buffer A, this treatment removed 79.83% of the non-PHB materials, and the PHB content in the final product was 90.53%.

Further purification of PHB was realized by a chemical treatment such as surfactant SDS treatment. As shown in figure 4, after treated by buffer A, the sediment with 90.53% PHB content in table 1 was treated by different concentration of SDS solution. When the concentration of SDS was 4g/L, the purity of PHB could be increased from 90.53% to 99.66%.

Table 1
Comparison of PHB content of *E. coli* JM109(pTU14) cells after buffer A treatment

PHB content of initial cells, P_{PHB0} (%)	PHB content after buffer A treatment, P_{PHB} (%)	Non-PHB materials extracted, θ (%)
41.10	69.03	68.69
43.23	71.72	69.95
46.29	74.70	70.81
60.61	85.15	73.17
66.07	90.53	79.63

*Initial cell mass: 8.5g/L. θ is defined as: $\theta = \frac{P_{PHB} - P_{PHB0}}{(1 - P_{PHB0})P_{PHB}} \times 100\%$

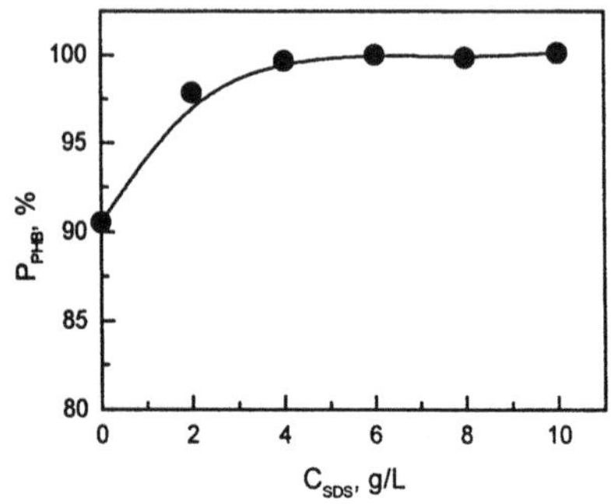

Figure 4. Purification of PHB by surfactant SDS treatment.

4. CONCLUSIONS

The one-plasmid system which contains the lambda (λ) phage lysis genes S(-)RRz and the biosynthesis genes of PHB is successfully developed and proved to be practicable for PHB producing. The combined biologi-chemical process for PHB separation from *E. coli* JM109(pTU14) involves three steps:

(1) Culture the *E. coli* JM109(pTU14) to accumulate PHB and R and Rz gene products in cells;

(2) Induce the *E. coli* JM109(pTU14) cells to lyse by buffer A treatment;

(3) Purify PHB granules from the lysed cells by surfactant SDS treatment.

The authors suggest that this coupling process have potential advantages to get high purity and high molecular weight of PHB products simply and cheaply.

REFERENCES

1. J.M. Ramsay, E. Berger, R. Voyer, *et al.*, Biotechnol. Tech., 8 (1994) 589.
2. E. Berger, B.A. Ramsay, J.A. Ramsay, *et al.*, Biotechnol. Tech., 3 (1989) 227.
3. P.A. Holmes and G.B. Lim, Separation Process, US Patent No. 4 910 145 (1990).
4. W. Lubitz, Lysis of bacteria for the release of poly-3-hydroxycarboxylic acids, Ger. Offen. DE 4 003 827 (1990).
5. D. Dennis, Method for the improved production and recovery of poly-β-hydroxybutyrate from transformed *Escherichia coli*, US Patent No. 5 512 456 (1996).
6. J. Garrett, R. Fusselman, J. Hise, *et al.*, Mol. Gen. Genet., 182 (1981) 326.
7. J. Sambrook, E.F. Fritsch and T. Maniatis (eds.), Molecular Cloning: A Laboratory Manual, 2nd edition, Cold Spring Harbor Laboratory Press, New York, 1989.
8. A.C. Ward and E.A. Dawes, Anal. Biochem., 52 (1973) 607.

Bioseparation Engineering
I. Endo, T. Nagamune, S. Katoh and T. Yonemoto (Editors)

Cleaning Liquid Consumption and Recycle of Biopharmaceutical Plant

S. Murakami[a], R. Haga[b] and S. Yamamoto[c]

[a]Kasado Material Handling & Industrial Plant Product Division, Hitachi Ltd.,
794 Higashitoyoi, Kudamatsu, Yamaguchi 744-8601, Japan

[b]Power & Industrial Systems R&D Laboratory, Hitachi Ltd.,
7-1-1 Omika, Hitachi, Ibaraki 319-1292, Japan

[c]Yamaguchi University,
2557 Tokiwadai, Ube, Yamaguchi 755-8611, Japan

In order to evaluate and reduce cleaning liquid consumption for biopharmaceutical plant, we have studied relations between structure and cleaning operation, on cleanability of piping and equipment by model cleaning experiments. Such parameters for cleanability include pipe liquid flow rate, liquid temperature, wall shearing force, and spray jet energy of a spray ball.
Keywords: pharmaceutical plant, CIP, cleaning, shearing force, spray ball

1. Introduction

Recently, cleaning of manufacturing facilities and production equipment is one of the most focused concerns in the pharmaceutical industry[1,2]. A cleaning system design which evaluates and assures an ability to adequately clean surfaces will not only help minimize the risk of contamination from impurities but also minimize cleaning liquid consumption and enable its recycle use by reduced residue concentration in the cleaning liquid.

The design of cleaning processes requires consideration of both equipment materials and methods of construction, equipment layout, process equipment design appropriate to cleaning in place, CIP, and finally the design of the manufacturing process[3-5].

At early stages of plant development, equipment structures satisfying impurity acceptance criteria are designed, and parameters for the cleaning operations are decided. Such parameters include time, temperature, flow rate, cleaning agent type, cleaning agent concentration, and so on. In order to establish a successful cleaning system at the design stage, some design bases are necessary. One must predict a cleanability of the equipment from its structure and operating condition for designing a pharmaceutical plant effectively and economically. Usually, design of each structure has been performed using experience design guidelines[6-9]. For example, it is said that process lines should be cleaned at a velocity ensuring turbulent flow in pipelines, which is usually achieved at a velocity of 1.5 (m/s)[6,9]. However, quantitative effect of the flow velocity on the cleanability, and applicability of this guideline is not well known. Therefore, pre-determined acceptance criteria for the cleanability have no direct effect on the equipment design through such experience guidelines.

We have studied relations between structure and cleaning operation, on cleanability of piping and equipment. Such physical parameters include surface finish, detergent temperature, liquid flow rate, spray ball jet energy, tank nozzle length, as well as time of the cleaning operation. Protein rich model residues were selected considering residues for bio-pharmaceutical plant.

2. Materials and Methods

2.1. Process Pipe

In order to evaluate the effects of pipe inner surface finish and temperature, dried milk on a plate of various surface finish was used as a model stain. A given amount of milk was dropped on the stainless steel plate of each surface finish (SUS304 stainless steel, 15×110 (mm)), and dried at 70 (℃) for 40 minutes. A sample plate with the model stain was suspended in an one liter beaker containing 1% caustic soda which was agitated by a magnetic stirrer at 100 rpm, as shown in Figure 1. Temperature was controlled between 30 to 70 (℃). The stain sample remained after the cleaning operation was ultrasonically dissolved into 0.1% caustic soda solution at ambient temperature. The dissolved solution was analyzed by calorimetric method (Protein-Assay Kit, Cat. No. 500-0006, Bio-Rad Laboratories, Hercules, CA.)

Effect of a flow rate on removal of a solid residue remaining in a pipe was investigated as follows. A plate with stain sample made by the same method above (SUS304 stainless steel, 2B surface finish, 15×110 (mm)) was set in a 1.5S sanitary straight pipe. Cleaning liquid, 1% caustic soda, was fed at a given flow rate, as shown in Figure 2. After pre-determined running time, the amount of stain remained was analyzed by the same method mentioned above.

2.2. Process Tank

In order to study spray ball cleaning which is often used for a process tank, an elemental model with single nozzle spray was prepared as shown in Figure 3. Liquid flow from the single nozzle hit the target plate, and separate itself into splashed flow and flow along the plate. The amount of non-splashed flow along the target plate was measured by collecting liquid falling down along the plate.

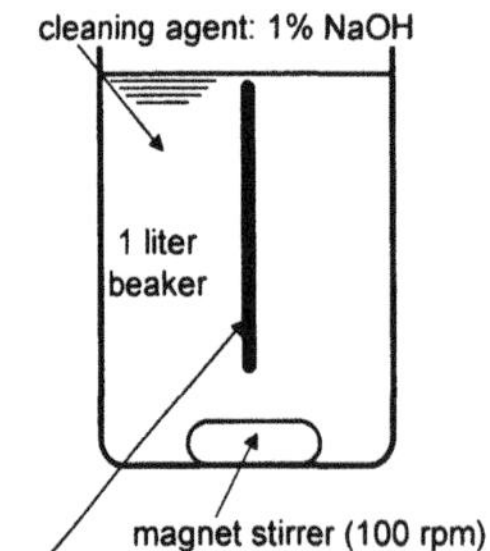

Figure 1. Pipe Surface Experimental Setup

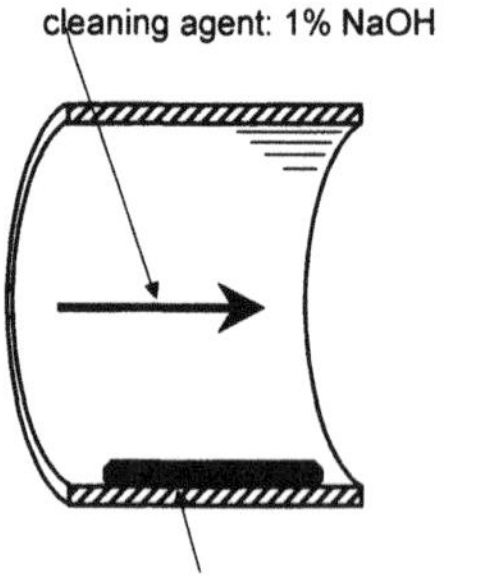

Figure 2. Effect of Flow Rate Experimental Setup

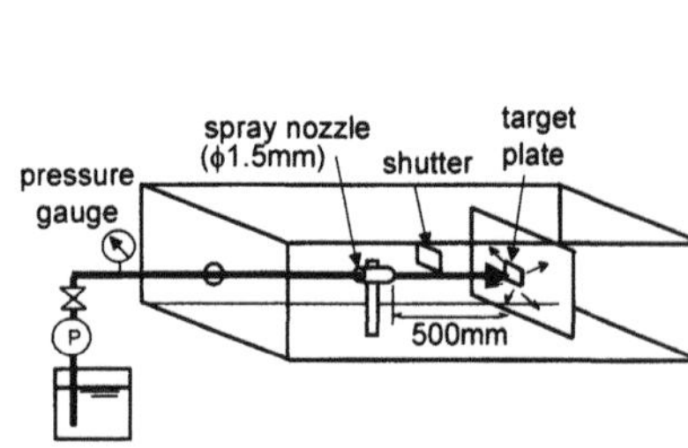

Figure 3. Spray Jet Experimental Setup

An upper part of a process tank exposed to a gas phase is often covered with a solid residue generated by powder material addition, aerosol formation or cultured bacterial growth. To investigate removal of solid residue, cheese paste, whip cream, cream cheese and butter were used as a model stain. Viscosity of the model stain was measured by a concentric cylinder rotating viscometer (PM-1B, Malcom, Ltd., Japan.) The model pastes were spread on the target plate (stainless steel, 2B surface finish, shown in Figure 3) in the area of 10 (cm^2) and thickness of 3 (mm). Tap water of 25 (℃) was used as cleaning liquid. After spray jet cleaning with predetermined time, the model paste was partly removed and base stainless steel was appeared as a round shape. These removal areas were measured as the spray jet cleaning abilities.

Finally, cleaning of a tank nozzle inner surface was simulated considering spray jet direction and nozzle length. Model nozzles with various size was aimed by the single jet of 28 (℃) tap water. Wetted heights were measured as spray jet cleaning abilities.

3. Results and Discussion

3.1. Process Pipe

(1) Surface Finish and Temperature

Between 30 to 70 (℃), the higher the temperature, the higher the residue removal speed was measured as shown in Figure 4. However, no significant difference has been observed among different surface finishes. Temperature dependence of the cleaning speed suggest its relation to chemical reaction, i.e. activate energy, rather than physical detachment force. Therefore, surface finish might not significantly affect the cleaning for this experiment. However, in spite of these experiment results, many other aspects should be considered when determining product contacting surface of pharmaceutical plant[8,10].

(2) Effect of Flow Rate

Because previous experiments were performed under relatively low fluid force in a stirred beaker, effect of higher fluid force was investigated in a straight pipe. Temperature of the fluid, tap water, was controlled between 30 to 70 (℃). As shown in Figure 5, residue removal speed was higher at high flow rates, i.e. high shearing stress, and high temperatures.

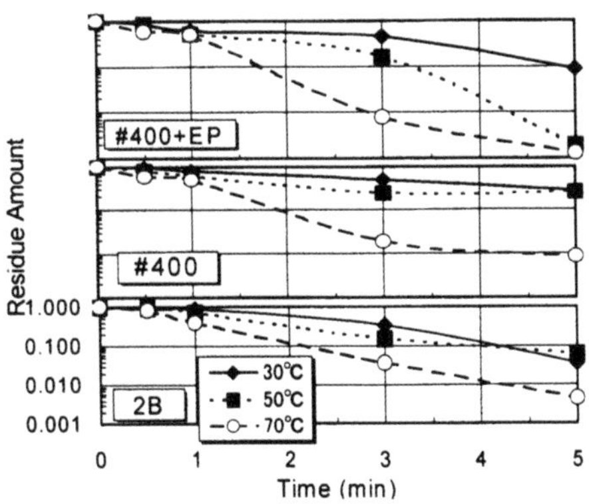

Figure 4. Pipe Surface and Temperature

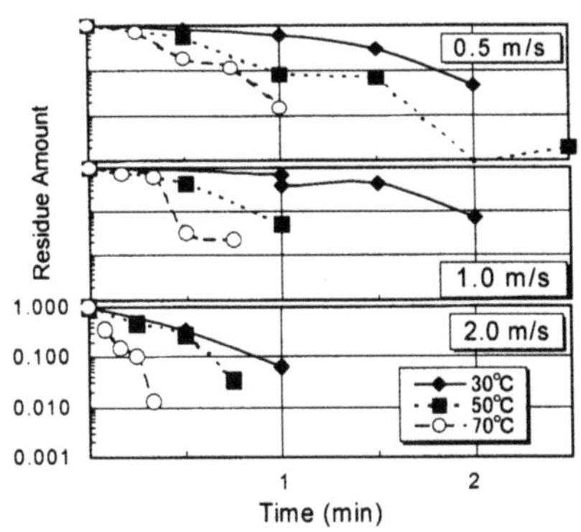

Figure 5. Effect of Flow Rate

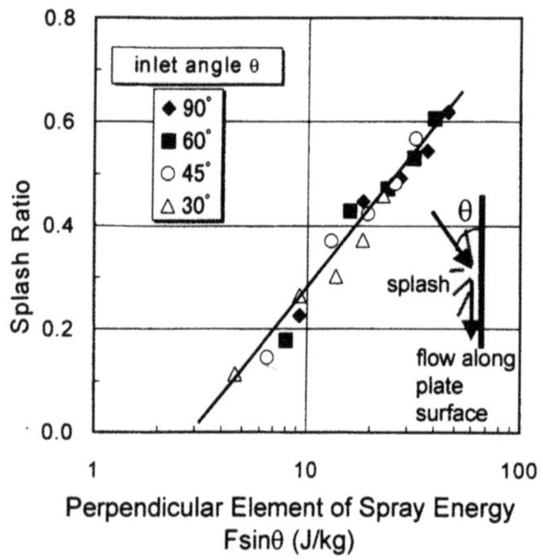

Figure 6. Splash Ratio and Spray Energy

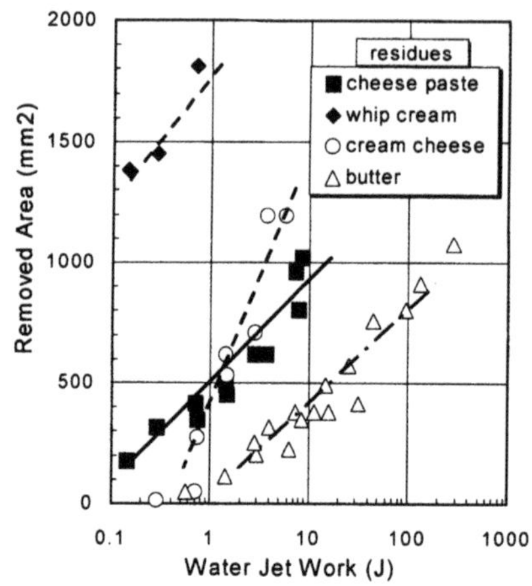

Figure 7. Solid Residue Removal

From this result, actual cleaning abilities of flow rates around usual experience guideline, 1.5 (m/s), can be predicted. In this experiment, residue removal was higher than the previous experiment of lower fluid force. Accordingly, not only chemical reaction but also fluid force may have acted as cleaning mechanism.

3.2. Process Tank

(1) Spray Jet Energy

During spray ball cleaning of a tank inside, a spray jet reaching the tank inner surface divides itself into two parts. One part splashes over inner space and detaches from the tank, and the other part flows along tank inner surface and cleans lower part of the tank, such as shell and bottom head. It is often said that spray ball flow rate should be 30 liter per minute for each meter of tank inner circumference[7]. However, a splashed jet will have no effect on lower wall surface cleaning, and only flow along tank inner surface will provide cleaning force.

Table.1
Model Residue for Spray Jet Cleaning Experiment

name	viscosity (Pa·s)	kinetic viscosity (m^2/s)	density (kg/m^3)	manufacturer
cheese paste	6.05	0.00585	1034	Koiwai Dairy Products Co.,Ltd., Japan
whip cream	1.36	0.00523	367	Nagoyaseiraku Co.,Ltd., Japan
cream cheese	11.3	0.0108	1042	Snow Brand Milk Products Co.,Ltd., Japan
butter	6.1	0.00633	963	Snow Brand Milk Products Co.,Ltd., Japan

Therefore, splash ratio, a proportion of splashed liquid to a total jet flow rate, should be considered in determining a spray ball flow rate. As shown in Figure 6, splash ratios are well correlated to perpendicular elements of spray energy, and this can be used for splash ratio prediction.

(2) Solid Residue Removal

Viscosity, μ, of the model stain was measured at 25 (℃). Physical properties of each model residue are shown in Table.1. Experimental results of model solid residue removal are shown in Figure 7. Water jet work, W, was calculated as follows:

$$W = \frac{1}{2} Q \rho v^2 t \qquad (1)$$

Where,
W: water jet work (J)
Q: liquid flow rate (m^3/s)
ρ: liquid density (kg/m^3)
v: liquid jet velocity (m/s)
t: operation time (s)

Consequently, necessary number of spray ball holes can be determined by dividing a tank head surface area by predicted removal area of single jet.

(3) Tank Nozzle

As shown in Figure 8, small nozzles less than 10 mm can be wetted as high as 600 mm. However, larger nozzles were difficult to be wetted. Therefore, tank nozzles larger than 15 (mm) should be kept shorter than 100 (mm) in order to supply enough cleaning liquid by spray ball.

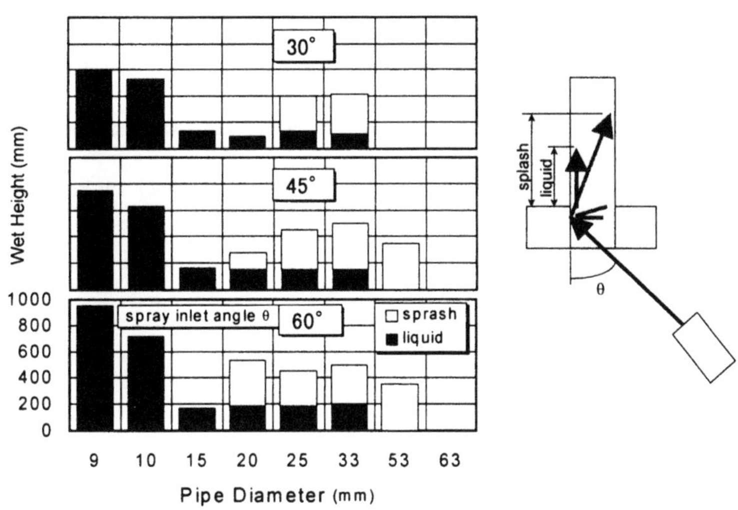

Figure 8. Tank Nozzle Cleaning

4. Conclusions

With model cleaning experiments, we have studied relations between structure and cleaning operation, on cleanability of piping and equipment.

i) With small sharing force, surface finish has negligible effect on solid residue removal comparing with temperature effect.

ii) Wall shearing stress is a major cleaning mechanism of solid residue removal at flow velocities around 1.5 (m/s).

iii) Splash ratios, which should be considered when calculating spray ball flow rate, are in proportion to perpendicular elements of spray energy.

iv) Solid residue removal area are in proportion to spray jet work.

v) Tank nozzles larger than 15 (mm) should be kept shorter than 100 (mm) in order to supply enough cleaning liquid by spray ball.

References

1. Food and Drug Administration, Guide to Inspections of Validation of Cleaning Processes, FDA, Rockville, MD, (1993).
2. Food and Drug Administration, International Conference on Harmonisation; Guideline on Impurities in New Drug Substances, Federal Register 94 D-0325, FDA, Rockville, MD, (1996).
3. T. Myers, T. Kasica and S. Chrai, Approaches to Cycle Development for Clean-in-Place Processes, *J. Parent. Sci. Technol.*, **41** (1987) 9.
4. H. Baseman, SIP/CIP Validation, *Pharm. Eng.*, **12** (1992) 37.
5. J. Agalloco, "Points to Consider" in the Validation of Equipment Cleaning Procedures, *J. Parent. Sci. Technol.*, **46** (1992) 163.
6. Y. Chisti and M. Moo-Young, Clean-in-Place Systems for Bioreactors: Design, Validation and Operation, *Bioprocess Engineering*, **27**, ASME, New York, NY, (1993) 5.
7. J. Stewart and D. Seiberling, Cleaning in Place, *Chem. Eng.*, **103**, January (1996) 72.
8. J. Voss (ed.), Cleaning and Cleaning Validation: A Biotechnology Perspective, Parenteral Drug Association, Bethesda, MD, (1996) pp. 3-40.
9. The American Society of Mechanical Engineers, An American National Standard, Bioprocessing Equipment, ASME, New York, NY, (1997).
10. J. Valley and L. Rathbun, Finishing Requirements for Stainless Steel in Parenteral Applications, *Bull. Parenter. Drug Assoc.*, **31** (1977) 94.

Index of Authors

www.ingramcontent.com/pod-product-compliance
Ingram Content Group UK Ltd.
Pitfield, Milton Keynes, MK11 3LW, UK
UKHW020809290926
15470UKWH00003B/1054

9 780444 504418